United States Nuclear Regulatory Commission

Protecting People and the Environment

NUREG/CR-7148
BNL-NUREG-98563-2012

I0482701

Confirmatory Battery Testing: The Use of Float Current Monitoring to Determine Battery State-of-Charge

Office of Nuclear Regulatory Research

AVAILABILITY OF REFERENCE MATERIALS
IN NRC PUBLICATIONS

NUREG/CR-7148
BNL-NUREG-98563-2012

United States Nuclear Regulatory Commission

Protecting People and the Environment

Confirmatory Battery Testing: The Use of Float Current Monitoring to Determine Battery State-of-Charge

Manuscript Completed: October 2012
Date Published: November 2012

Prepared by:
W. Gunther, G. Greene, M. Villaran, Y. Celebi,
and J. Higgins

Brookhaven National Laboratory
Nuclear Science and Technology Department
Systems Engineering Group
Upton, NY 11973-5000

Liliana Ramadan, NRC Project Manager

NRC Job Code N6542

Office of Nuclear Regulatory Research

ABSTRACT

In February 2007, the U.S. Nuclear Regulatory Commission (NRC) issued Regulatory Guide (RG) 1.129 Rev. 2, "Maintenance, Testing, and Replacement of Vented Lead-Acid Storage Batteries for Nuclear Power Plants." In this RG, the NRC staff endorsed the Institute of Electrical and Electronics Engineers (IEEE) Standard 450-2002, "Recommended Practice for Maintenance, Testing, and Replacement of Vented Lead-Acid Batteries for Stationary Applications." This standard provides the recommended practices, test schedules, and testing procedures including recommended methods for determining a battery's state-of-charge to maintain permanently installed vented lead-acid storage batteries (typically of the lead-calcium type) for their standby power applications. Previous versions of this standard suggested that either float current or specific gravity could be used for determining the battery's state-of-charge. The NRC sponsored the research project described herein to validate the use of float charging current as a measure of a battery's state-of-charge for batteries that are used in the nuclear industry. This report describes the approach taken, the specific activities performed to achieve the objectives of this research effort, and the results achieved. It provides analysis of the data and offers observations and recommendations for use by the NRC and its licensees.

TABLE OF CONTENTS

LIST OF FIGURES

LIST OF TABLES

EXECUTIVE SUMMARY

To ensure that a battery has the capability to execute its safety function, it is necessary to confirm its fully charged condition and operational readiness. For the past three decades the typical nuclear power plant Technical Specifications required the measurement of specific gravity to determine the state-of-charge of the batteries. This requirement was based on RG 1.129 Rev.1, "Maintenance, Testing, and Replacement of Large Lead Storage Batteries for Nuclear Power Plants," which endorsed IEEE Std. 450-1975, IEEE Recommended Practice for Maintenance, Testing, and Replacement of Large Lead Storage Batteries for Generating Stations and Substations." The more recent version of this standard, IEEE Std. 450-2002, that was endorsed by the NRC, suggested that either float current or specific gravity could be used for determining a vented lead-calcium battery's state-of-charge. This report describes the project to validate this approach on batteries that are used in the nuclear industry.

In conducting this study, three sets of nuclear qualified batteries were procured from three different battery vendors. Each battery set consisted of 12 battery cells. These cells are the same models that are typically used in a Class IE dc system application. Two suitably sized battery chargers and a load bank were also obtained; the second battery charger was used to maintain the batteries not being tested on a continuous float charge. The test setup was as close as reasonably practicable (not including seismic battery racks) to a typical nuclear power station's Class 1E battery design. Once the battery was fully charged and stabilized, a 4-hour discharge test was performed based on the battery vendor's specifications. After the discharge test, the float current was continuously recorded while the battery was recharged. During the recharge, periodic specific gravity measurements were taken at the vendor-specified location (about 1/3 down the length of the cell) for all of the cells, and at the top and bottom as well for two of the cells. The three measurements taken on two of the cells allowed us to obtain a vertical profile representation of the electrolyte's distribution within the cell during the entire discharge-recharge cycle. Discharge test current and specific gravity readings were compensated for temperature as discussed in IEEE Std. 450-2002.

The major findings that were derived from more than thirty cycles of deep discharge testing are:

1) Both float current and specific gravity provide adequate means to determine battery state-of-charge. Float current has an advantage in that it provides an indicator of the entire battery string, while specific gravity is measured on a cell by cell basis.

2) Both float current and specific gravity have similar response times when the battery is recharged. Generally speaking, 100% of the ampere-hours discharged are returned to the battery within 24 hours of the start of the recharge cycle.

3) The amount of electrolyte stratification is significant following a performance test and it takes months before equilibrium is reached within the cells. Therefore, it is critical to measure specific gravity at the correct point as indicated by the battery vendor's manual and supported by IEEE Std. 450-2002.

4) The use of pilot cells to ascertain specific gravity is supported by the consistent response observed among all cells during both discharge and recharge.

5) Measuring float current through the use of a simple shunt connected to a data acquisition system provides accurate and repeatable measurements. We used a 200 amp (A), 50 millivolt (mV) shunt placed in series with the output of the battery charger. We found that a more sophisticated device based on the principles of the Hall Effect

(similar to a clamp-on ammeter) is also effective, but it is less accurate at the low ends of the float current range (< 2 amps).

Other observations derived from the extensive testing that was performed are:

Using a standard length of tubing to draw the electrolyte from the same point resulted in consistent "trendable" specific gravity data.

- Temperature compensation for capacity testing, specific gravity readings and conductance readings is important. If not performed properly, the data will be skewed.

- Float current response will vary based on the recharge voltage applied to the battery. However, regardless of the voltage applied during recharge, the float current of a nearly fully charged battery becomes stable at less than two amps.

- Calculations of the ampere-hours returned to the battery during recharge can be used to verify the battery's state-of-charge. The majority (>60%) of the ampere-hours returned to the battery occurs while the battery charger is still in a current limit mode.

IEEE Std. 450-2002 contains the following criterion related to return to service for a battery: "When the charging current has stabilized at the charging voltage for three consecutive hourly measurements, the battery is near full charge." Our test program also verified the point where the battery can be safely returned to service. In a series of six additional tests (two tests per battery string), the battery strings were able to meet their capacity and capability requirements at the point where the float current was stable for three hours. Thus the criterion used in IEEE Std. 450-2002 was found to be an acceptable practice for ensuring the capacity and capability requirements of the battery were met before returning it to service.

Similarly, three cycles of tests were performed in which each battery was returned to service when the float current reached the value equivalent to three time constants on the recharge/float current curve. This occurred within about twelve hours and at a higher current than the previously described return to service tests. In each case, the battery was also able to meet its capacity and capability requirements. This calculated float current value obtained from the battery-specific recharge/float current curve may be a more practical method for returning the battery to service at the point where it is capable of meeting its capacity and capability requirements.

ACKNOWLEDGMENTS

The authors wish to thank Liliana Ramadan, NRC Project Manager, for her guidance and input during the development, conduct of testing and writing of this report. We thank her NRC colleagues Thomas Koshy and Matthew McConnell for their insights and helpful comments on the test program. We appreciate the support provided by Robert Lofaro and Edward Grove of BNL in the development of the project management and quality assurance plans, respectively. We also wish to thank William Maloney, formerly of BNL, for his input during the preparation of the testing program and Paul Giannotti of BNL, for his technical expertise in developing some of the support instrumentation. We acknowledge and appreciate the administrative support from Maryann Julian in producing this document.

1. INTRODUCTION

1.1. Background

In February 2007, the U.S. Nuclear Regulatory Commission (NRC) issued Regulatory Guide (RG) 1.129 Rev. 2, "Maintenance, Testing, and Replacement of Vented Lead-Acid Storage Batteries for Nuclear Power Plants." In this RG, the NRC staff endorsed the Institute of Electrical and Electronics Engineers (IEEE) Standard 450-2002, "Recommended Practice for Maintenance, Testing, and Replacement of Vented Lead-Acid Batteries for Stationary Applications." This standard provides the recommended practices, test schedules, and testing procedures including recommended methods for determining a battery's state-of-charge to maintain permanently installed vented lead-acid storage batteries (typically of the lead-calcium type) for their standby power applications.

To ensure that a battery has the capacity and capability to execute its safety function, it is necessary to confirm its fully charged condition and operational readiness. For the past three decades, the typical nuclear power plant Technical Specifications required the measurement of specific gravity to determine the state-of-charge of the batteries. This requirement was based on Regulatory Guide (RG) 1.129 Rev.1, "Maintenance, Testing, and Replacement of Large Lead Storage Batteries for Nuclear Power Plants," which endorsed IEEE Std. 450-1975, "IEEE Recommended Practice for Maintenance, Testing, and Replacement of Large Lead Storage Batteries for Generating Stations and Substations." A more recent version of this standard, IEEE Std. 450-2002, that was endorsed by the NRC in Rev. 2 of RG 1.129, suggested that either float charging current or specific gravity could be used for determining a vented lead-calcium battery's state-of-charge. The testing program approved by the NRC using a series of 4-hour performance tests was implemented to validate this approach on batteries that are commonly used in the nuclear industry. Comparisons were made of the recharge/float current and the specific gravity responses as the cells were charged following the four hour performance test. Note that these test results are only applicable to vented lead-calcium batteries.

1.2. Program Objectives

The primary objective of this research project was to determine whether the float charging current can be a useful indicator for determining a vented lead-calcium battery's state-of-charge over the life of the battery. This project evaluated the acceptability of using float charging current as a means of monitoring battery state of charge for lead-acid calcium batteries from three vendors. A secondary objective was to evaluate the point at which a battery could be returned to service and meet its performance requirements.

1.3. Research Approach

The approach taken in this research project involved testing of Class 1E batteries representative of those used in commercial nuclear power plants. These batteries were installed in a configuration similar to that used in the nuclear power plants and were subjected to full discharge and recharge cycling equivalent to what would be experienced over its nominal 20-year life. The number of cells used in the testing (12-cells per battery string) is smaller than what is typically used in a nuclear power plant, where 60 or 120 cell batteries are employed. However, the test results of the 12-cell string are directly applicable to the larger batteries since they are carrying the same current and the cell voltages are the same. The scaling of the overall battery voltage is linear to the number of cells in the string. Note that this testing did <u>not</u>

age the batteries by subjecting the batteries to elevated temperatures. Charging (float) current and specific gravity were measured during each test cycle in accordance with IEEE Std. 450-2002, along with other battery parameters, to monitor the status of the batteries. Testing was performed in accordance with a Quality Assurance Plan developed specifically to meet the needs of this project that ensured an acceptable level of quality for the test results.

1.3.1. Establish Test Facility

BNL established a controlled area for this testing that achieved the needed environmental and electrical safety parameters contained in IEEE and manufacturer's standards, and met BNL safety procedures. Facility attributes included area temperature and humidity control and monitoring, electrolyte spill control measures, and adequate ventilation to prevent hydrogen accumulation. Security measures were established so that access to the testing area was limited to those directly involved with the battery confirmatory testing program. Data acquisition equipment was installed to acquire and store the measured parameters during testing.

1.3.2. Laboratory Space

Testing was performed in a high bay area located in Building 526 at BNL. The dedicated space for the battery testing is approximately 800 square feet, has a controlled heating and ventilation system and adequate electrical power sources to support the battery charging and associated test equipment. In addition, the space was upgraded in accordance with battery vendor recommendations to include a containment system to capture any spilled electrolyte from the cells and a hydrogen monitor to detect any accumulation of explosive gasses in the test area.

1.3.3. Battery Test Setup

Figure 1-1   Battery test laboratory at BNL

Battery racks were used that replicate a nuclear power plant installation with the exception that the racks were not seismically qualified. There are three racks of 12-cells each representing the three battery vendors. Adequate spacing between the racks was provided to ensure that test leads could be attached and specific gravity and conductance measurements could be safely taken for all of the cells. The installation is shown in Figure 1-1.

1.3.4. Data Acquisition Equipment

Nuclear power plants employ test equipment for the performance tests required by their technical specifications. One of the most common units used is the BCT-128™ manufactured by Alber Corp (Figure 1-2). This test equipment monitors and displays cell voltages and can be programmed to discharge a battery under constant power, constant current, or variable current. For the purposes of our testing, we employed the test set in conjunction with a load bank, to discharge the battery at a constant current for a four hour performance test. The unit automatically disconnected the load when the overall string voltage reached 21.0 volts. It was also configured to shut down the test if any one cell reached 1.6 volts; however, that never occurred.

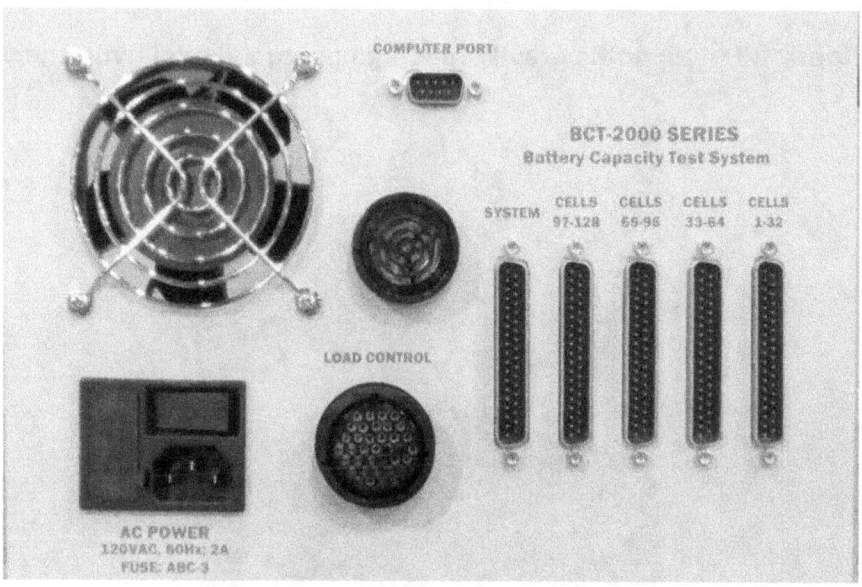

Figure 1-2   Battery capacity test set (Alber BCT-128™)

In addition to the use of test equipment that acquires a continuous stream of data during testing, several manual operations were required periodically. These included specific gravity readings, intercell resistance readings, and conductance readings. The intercell resistance readings ensured the integrity of the connections; the conductance readings were taken to determine if they could provide an indicator of cell state-of-charge.

Test equipment traceable to the National Institutes of Science and Technology (NIST) was used to acquire these data. Figure 1-3 illustrates the use of a Storage Battery Systems SBS-2500 digital hydrometer for the required specific gravity readings, and Figure 1-4 shows the use of the Midtronics Celltron Ultra CTU-6000 universal stationary battery analyzer to obtain conductance measurements.

3

Figure 1-3   Specific gravity readings using a digital hydrometer

Figure 1-4   Conductance readings in progress

1.3.5.   Research Project Management

Discharge (performance) and recharge testing were performed on only one battery string at a time to ensure that proper controls and monitoring were focused on that activity. We established a test schedule that resulted in one test cycle being completed per week. This allowed the battery string to be in a float mode with stable float current for at least 120 hours before subjecting the battery to another performance test.

The batteries that were not actively undergoing testing were maintained on float charge using a separate battery charger. Periodic inspection and maintenance of the batteries not under active testing was performed to maintain them in good condition and a fully charged state.

Table 1-1 shows the activities performed along with the scheduled milestones.

Table 1-1   Confirmatory Testing Research Project Schedule.

Activity	FY09 Qtr. 4	FY10 Qtr. 1	Qtr. 2	Qtr. 3	Qtr. 4	FY11 Qtr. 1	Qtr. 2	Qtr. 3	Qtr. 4	FY12 Qtr. 1-2
Develop Plans		X								
Program Test Plan		X								
Quality Plan		X								
Procurement Plan		X								
Milestone: Plans Complete (Draft)		X								
Milestone: Plans Complete (Final)			X							
Establish Test Facility	X	X	X							
Identify suitable space for testing	X									
Set up laboratory space		X	X							
Procure and install data acquisition equipment			X	X	X					
Perform safety review and obtain approval to test					X					
Milestone: Test Facility Ready					X					
Acquire Test Specimens		X	X	X	X	X				
Order battery, charger, rack, and load bank- all vendors		X								
Receive and install chargers, battery racks, & load bank			X							
Vendor 1 installs batteries					X					
Vendor 2 installs batteries					X					
Vendor 3 installs batteries						X				
Milestone: All Test Specimens Acquired and Installed						X				
Perform Battery Float Current Tests					X	X	X			
Milestone: Complete Vendor 1 Testing – 10 cycles						X				
Milestone: Complete Vendor 2 Testing – 10 cycles							X			
Milestone: Complete Vendor 3 Testing – 10 cycles							X			
Perform Return to Service Battery Tests									X	X
Complete stable float current testing									X	
Complete 3 time constant testing										X
Milestone: Complete Return to Service Testing										X

2. TESTING PROTOCOLS

2.1. Introduction

Testing of the first battery string, the Enersys Model 2GN-23, was initiated on September 21, 2010 with a four-hour performance test (discharge test). Following successful completion of the performance test, the battery was recharged for 120 hours. Prior to the start of the discharge test, baseline readings of specific gravity, cell conductance, and cell temperature were taken. During the recharge cycle, measurements of specific gravity, cell voltage, cell temperature, cell conductance and float current were completed.

The Alber Battery Capacity Test Set was used to control and monitor the performance test. It provided data on the discharge current, individual cell voltages, and a real-time calculation of battery capacity. An Omega temperature data acquisition system was used to monitor cell temperatures over the entire cycle. A second Omega module was used to record the voltage across a calibrated shunt installed in series with the output of the battery charger. This shunt provided the float charging current output. A second means of measuring float charging current employed the Polytronics float current monitor. This instrument provided continuous readings of float current from the battery charger during the entire recharge cycle as well as individual cell voltages. Manual readings of specific gravity using an SBS-2500 digital hydrometer were taken periodically during the discharge-recharge cycle along with cell conductance readings using a Midtronics CTU-6000 conductance meter. The overall test process is illustrated in Figures 2-1 (discharge mode) and 2-2 (recharge mode).

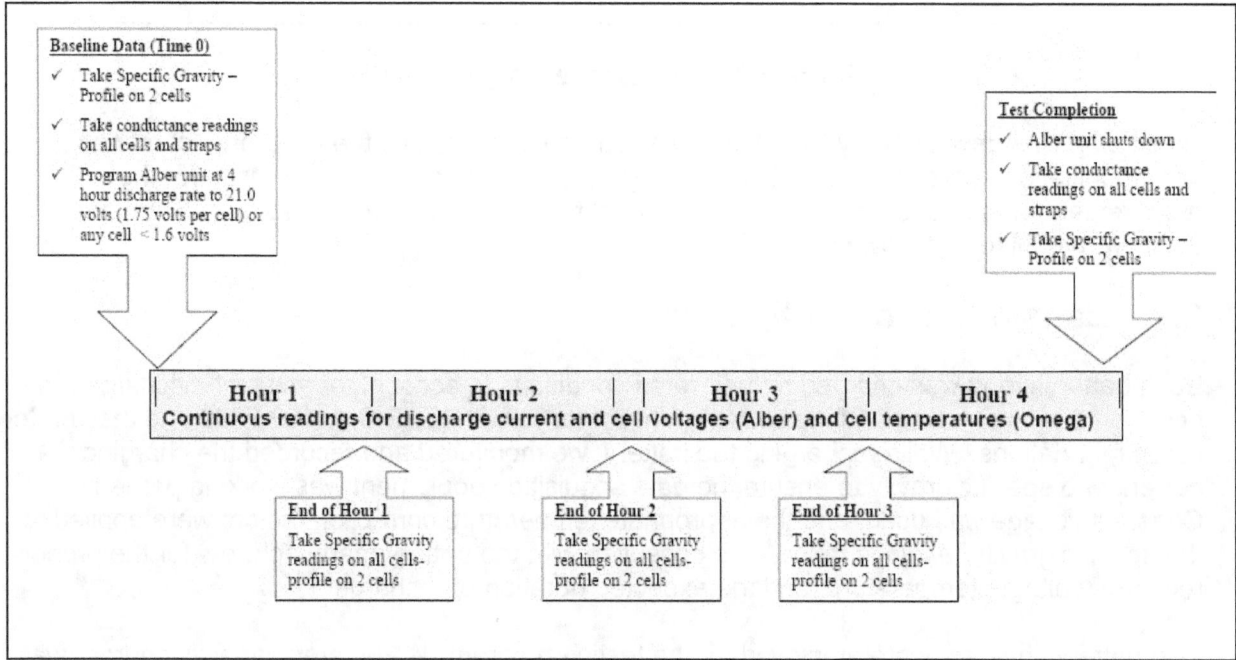

Figure 2-1   Battery discharge (performance) test

The same process was used for each of ten discharge/recharge cycles for the Enersys battery string, and then continued for the Exide GNB and C&D battery strings over 30 weeks of testing. Weekly, a summary level data report was provided to the NRC project manager followed up periodically with conference calls to discuss the data and its interpretation.

7

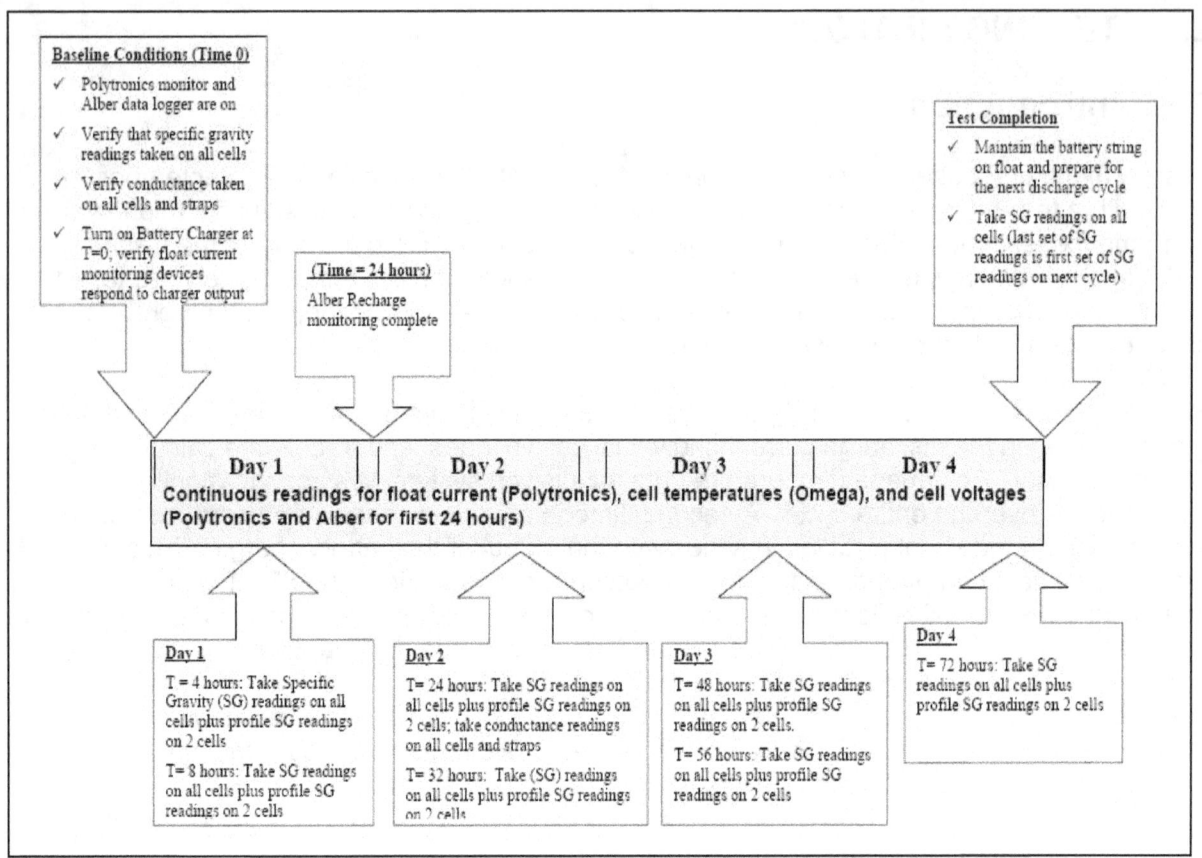

Figure 2-2   Battery recharge activities

A test plan was developed by BNL and reviewed and approved by the NRC project manager prior to initiation of the testing. This overarching plan included the process used for the procurement of the batteries, the management of the project, a quality assurance plan, and the testing protocol to be followed.

2.2. Battery String Configuration

Each battery string was charged to their rated conditions to account for any self-discharge that occurred during shipping and installation in accordance with IEEE Std. 450-2002 and the vendor recommendations. While recharging the battery, we monitored and recorded the charging current and specific gravity to ensure the data acquisition equipment was working properly. Constant voltage was used, and the appropriate temperature correction factors were applied to the specific gravity readings taken. We consulted with the battery manufacturers for the proper recharge voltage, temperature, and the expected duration of recharge.

Two battery chargers were employed for the testing program. A 200 amp capacity charger was used for recharging the battery string under test. This charger, manufactured by Enersys (Model AT30-24-200), has an output voltage regulation of 0.25% and an adjustable current limit mode that was set to 90% of rating or 180 amps. A second battery charger (100 amp capacity) was used to maintain the other two battery strings in a fully charged condition (float voltage of 27.0 volts). For the performance test, the charger was disconnected from the battery and a full-capacity discharge of the battery performed using the supplied load bank and the respective

8

manufacturer's specifications for a four-hour test. The discharge was automatically terminated when the battery string of 12-cells reached 21.0 volts (average of 1.75 volts per cell). A second automatic shutdown was programmed into the test unit if any cell reached a terminal voltage of 1.60 volts. This latter condition was never experienced during the testing program. Test parameters monitored during the deep discharge included cell temperatures, ambient area temperature, discharge current, terminal voltages, and periodic specific gravity and conductance readings.

The charger was reconnected and placed in a constant voltage mode (current limiting) to recharge the battery bank following completion of the performance test. The charging current was recorded at a high resolution during this iteration (milliamp level). Periodic midpoint specific gravity measurements were taken on all of the cells and specific gravity profiles (top, bottom and midpoint) were taken on two cells during discharge and recharge. As a matter of record, recharging of the battery commenced within about an hour following the discharge cycle in accordance with the vendor's recommendations. The midpoint reading was typically about 1/3 the distance from the top of the cell as specified by the battery vendor and supported by IEEE Std. 450-2002.

Ten charge and discharge cycles were conducted for each battery string. For the Enersys battery string, recharge was conducted at the high end of the float voltage range (27.0 volts) as recommended by the manufacturer. For the GNB and C&D batteries, recharge was conducted at the equalizing voltage of 28.0 volts on several cycles to compare the float current characteristics with the standard recharge voltage of 27.0 volts. This approach was approved by representatives from these battery vendors.

2.3. Test Equipment Used

Nuclear power plants employ test equipment for the performance tests required by their technical specifications. One of the most common units used is the BCT-128™ manufactured by Alber Corp (Figure 1-2). This test equipment monitors and displays cell voltages and can be programmed to discharge a battery under constant power, constant current, or variable current.

In addition to the use of test equipment that acquires a continuous stream of data during testing, several manual operations were required periodically. These included specific gravity readings, intercell resistance readings, and conductance readings.

Test equipment calibrated to standards traceable to NIST was used to acquire these data so that they can be correlated to changes in battery capacity. A float current monitoring device was used to delineate the charging current measured during testing of the first battery string and then was used as a backup for the remainder of the testing program. Figure 2-3 is a representation of the circuit containing the 200 amp, 50 mV calibrated current shunt that was used to determine the float current on the GNB and C&D battery strings and for all three battery strings during the "return-to-service" testing. Figure 2-4 is the 600 amp, 24 volts capacity load bank used in the testing.

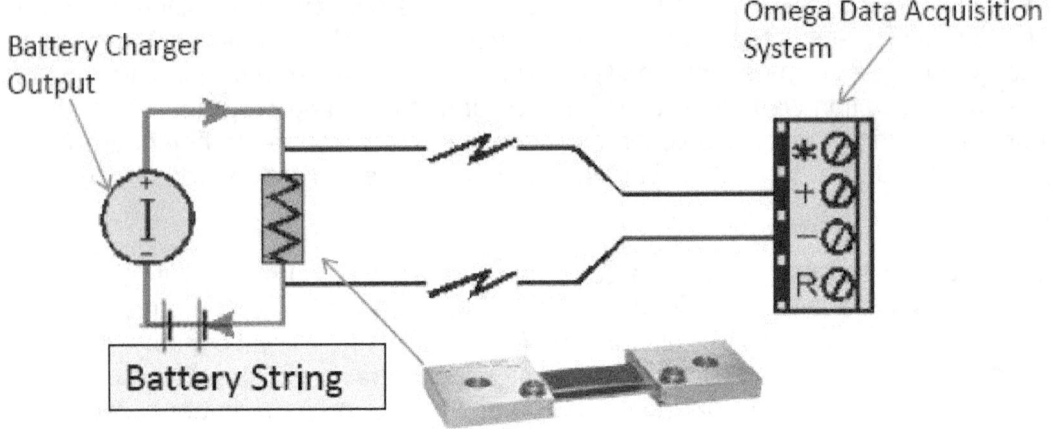

Battery Charger Output

Omega Data Acquisition System

Battery String

200A-50mV shunt

Figure 2-3   Circuit Using Calibrated 200 amp - 50 millivolt shunt

Figure 2-4   600 amp capacity resistive load bank

2.4. Test Cycle Description

The NRC approved test plan ensured that each battery string was tested in accordance with the manufacturer's requirements and that the testing was carried out consistently throughout the testing program. The plan also addressed the monitoring techniques to be evaluated and provided guidance on the steps to take if a failure occurred. Specific BNL procedures were used to augment the test plan. They provided precautions and prerequisites for conducting the test as well as detailed information about the data to be acquired.

The tests are summarized as follows:

Table 2-1   Test Activities.

Test Activity	Specimens Tested	Objective of Test
Vendor Acceptance Test (performed by vendor prior to shipment or at BNL following installation).	ALL	As part of the nuclear safety Class 1E procurement and installation, each battery will pass an acceptance test as required by IEEE Std. 450-2002.
Capacity Test-Time adjusted method	ALL-multiple times	This testing was conducted in accordance with section 7.4 of IEEE Std. 450-2002. The vendor's battery ratings were used to determine the loading to be applied. Parameters were monitored to determine changes that occur during deep discharges.
Charging	ALL-multiple times	Constant voltage, current- limited recharging of the battery string will be applied as soon as possible following the capacity test. Charging current was closely monitored and was used to determine when the battery string had reached full capacity. This is the key activity where specific gravity and charging current are compared to determine how accurately they reflect battery state-of-charge.
Cycle 11 Capacity Test was conducted after the electrolyte had reached an equilibrium state (specific gravity uniform in pilot cells).	All-one cycle each	Determine the impact of the stratified electrolyte on battery capacity.
Return-to-Service Testing: this series of tests were conducted several months after the completion of the first portion of the test to address a question regarding the point at which a battery can be returned to service following maintenance or discharge testing.	All-three cycles each (cycles 12-14 for each battery string)	Determine the point at which the battery can be returned to service with assurance that it can meet its capacity and capability objectives. Two methods tested-stable float current for three hours and three time constants.

2.4.1.   Three Types of Batteries in this Program

The specific batteries to be tested were selected by the NRC and are representative of the batteries used in most U.S. nuclear power plants. The specific lead-calcium batteries used were:

- Enersys 2GN-23 cells with a nominal capacity (8-hour rating) of 1800 Amp-hours,

- Exide GNB NCN-21 cells with a nominal capacity (8-hour rating) of 1496 Amp-hours, and

- C&D Technologies LCR-33 cells with a nominal capacity (8-hour rating) of 2320 Amp-hours.

The Exide GNB specimens were supplied by Nuclear Logistics Inc., the sole supplier of nuclear grade batteries for Exide.

2.4.2. Installation Processes

Qualified vendors were contracted to install the batteries at BNL. A baseline record of the mechanical and electrical installation parameters was documented. This included open circuit voltage, initial charge, and charging current readings. Connection resistances were measured to ensure the integrity of the connections.

2.5. Testing Process: Evaluation of Charging Current as a Monitoring Technique

The testing was conducted as delineated in Section 7.4 of IEEE Std. 450-2002 as follows:
- Set up the load and the necessary instrumentation to maintain the test discharge rate for battery string being tested. The discharge rate is based on the 4-hour battery rating.

- Disconnect the charging source, connect the load bank to the battery, start the timing, and continue to maintain the selected discharge rate. (This was accomplished using the Alber BCT-128 Capacity Test Set.)

- Maintain the discharge rate until the battery terminal voltage decreases to a value equal to the minimum average voltage per cell as specified by the design of the installation times the number of cells. For acceptance and performance tests as an example, a 12-cell battery with a minimum design voltage of 1.75 volts per cell, then the minimum battery voltage for the test is 12 × 1.75 or 21 volts.

- Read and record the individual cell voltages and the battery terminal voltage. These measurements were taken continuously by the data acquisition system. Other parameters such as specific gravity, intercell resistance, and conductance were taken while the load is applied at the beginning of the test, at regular intervals, and at the completion of the test. The process for taking and recording the parameter values are addressed in separate BNL procedures that follow good testing practices and vendor recommendations.

  Note: For safety reasons (concern for cell voltage reversal), the capacity test set was programmed to automatically terminate the test if any one cell reached 1.6 volts.

- Observe the battery for any abnormal conditions especially any intercell connector heating that could result in damage to the cell.

- At the conclusion of the discharge, the Alber battery capacity test set calculates the battery capacity in accordance with Section 7.3 of IEEE Std. 450-2002 applying the temperature corrections presented in Table 1 of IEEE Std. 450-2002 (see Table 2-2 in this document).

- The load bank is then turned off and the battery realigned to the battery charger. (Before energizing the charger, we conducted a series of specific gravity and conductance readings.)

While recharging the battery, we monitored and recorded the charging current and specific gravity. Constant voltage was applied, and the appropriate temperature correction factors made. Little change in electrolyte level occurred during the testing program so no electrolyte or distilled water needed to be added over the 10 cycles.

Table 2-3 summarizes the parameters that were measured during the conduct of the discharge and recharge cycles. Table 2-4 identifies the primary instrument used to measure the parameter and its precision.

Table 2-2   Recommended time correction factor (Table 1 of IEEE Std. 450-2002).

Initial temperature (°C)	Temperature correction factor K_T	Initial temperature (°C)	Temperature correction factor K_T	Initial temperature (°C)	Temperature correction factor K_T
5	0.684	22	0.966	30	1.045
10	0.790	23	0.977	31	1.054
15	0.873	24	0.986	32	1.063
16	0.888	25	1.000	33	1.072
17	0.902	26	1.006	34	1.081
18	0.916	27	1.015	35	1.090
19	0.929	28	1.025	40	1.134
20	0.942	29	1.036	45	1.177
21	0.954				

NOTE—This table is based on nominal 1.215 specific gravity cells. For cells with other specific gravities, refer to the manufacturer. Manufacturers recommend that battery testing be performed between 18°C and 32°C. These values are average for all time rates between 1 hour and 8 hours. See Annex L for the Fahrenheit conversion for Table 1.

Table 2-3   Parameters monitored over time.

Cycle #	Charging Current	Specific Gravity	Conductance	Intercell Resistance	Cell Voltage	Cell Condition
Discharge Cycle Charge Cycle	• N/A • Charger output to .01A	All cells at the midpoint level and profiles (top, midpoint, and bottom) of 2 cells at regular intervals	All cells prior to charge and discharge cycles	All cells prior to charge and discharge cycles	All cells continuously	Visual assessment Recorded following each cycle
Repeat ~10X						

Table 2-4   Summary of parameters measured.

Parameter	Test Equipment	Precision
Cell Voltage	Alber BCT-128	.01V
Inter-cell voltage	Midtronics CTU-6000	.001V
String volts & amps	Alber BCT-128	0.2V/0.1%A
Electrolyte Temperature	SBS-2500 Digital hydrometer	0.2° F
Float Current	Polytronics (for Enersys Battery) Calibrated Shunt for GNB and C&D	10 mA 1 mA
Cell Conductance	Midtronics CTU-6000	2%
Specific Gravity	SBS-2500 Digital hydrometer	0.0001 g/cm^3
Ambient Temperature	Omega thermocouple	0.5°F
Cell Surface Temperature	Omega thermocouple	0.5°F

3. TEST RESULTS

This section of the report discusses some of the data that were acquired using examples of representative test cycles. A complete set of data for every test cycle on each battery string is on file and maintained at BNL. Since the data taken on all three battery strings are fairly uniform, in some cases only data for one or two of the battery vendors are presented to illustrate the types of measurements taken and the responses obtained.

3.1. Performance Tests by Battery (Cycles 1-10)

3.1.1. Enersys Battery String

Cycle 1 Discharge Voltage Profile (Figure 3-1): Note that the starting voltage is 27.0 volts with the battery charger secured. When the discharge test is initiated, the battery voltage drops precipitously from 27.0 volts to about 23 volts. This same general behavior was experienced on all three batteries during each of the discharge tests. The rated load current for the Enersys battery is 385 amps corrected to 376 amps in this test to account for the electrolyte temperature of 73° F. (Performance test results are referenced to 77 °F per IEEE Std. 450-2002)

A resistive load bank was used and was controlled by the Alber Capacity Test Set at a steady current over the nominal four-hour test. The performance test was automatically terminated when the battery string voltage reached 21.0 volts. Note that "Battery OV" represents the overall battery string voltage (left axis), and "Battery Load" represents the discharge current to the load bank in amps (right axis). For this test, the automatic shutdown occurred in four hours and four minutes which translated to a capacity of 101.7%.

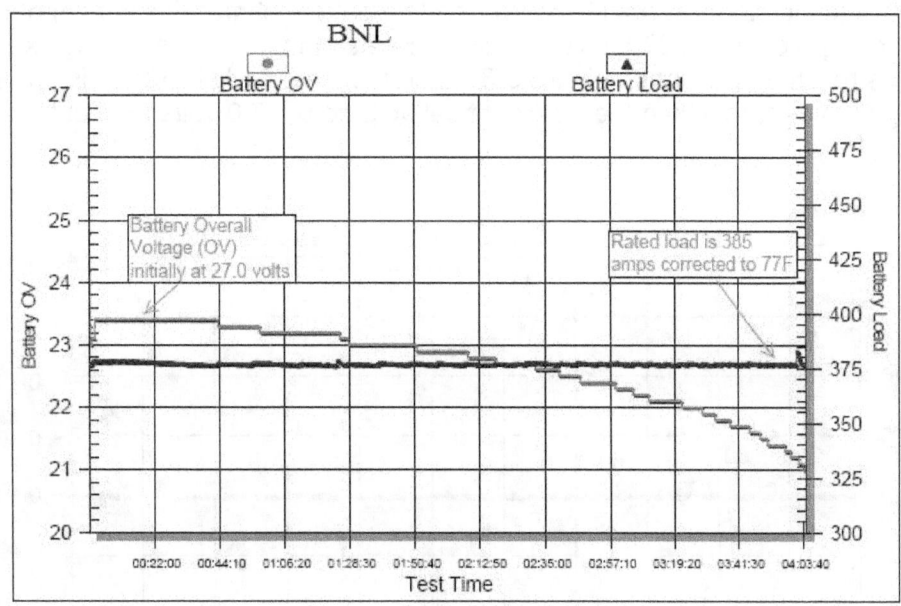

Figure 3-1   Enersys cycle 1 discharge voltage profile

The individual cell voltages were also monitored during the performance test and were observed to behave uniformly over the 10 test cycles within tenths of a volt. An example of one cell's voltage response during the cycle 1 performance test is provided in Figure 3-2. Similar voltage data for each individual cell were recorded that manifested themselves as a proportional

15

reflection of the overall battery string voltage versus time during the discharge and recharge cycles. During the conduct of the test, the 12-cell voltages were displayed on the computer monitor allowing us to compare and measure individual cell voltages in real time.

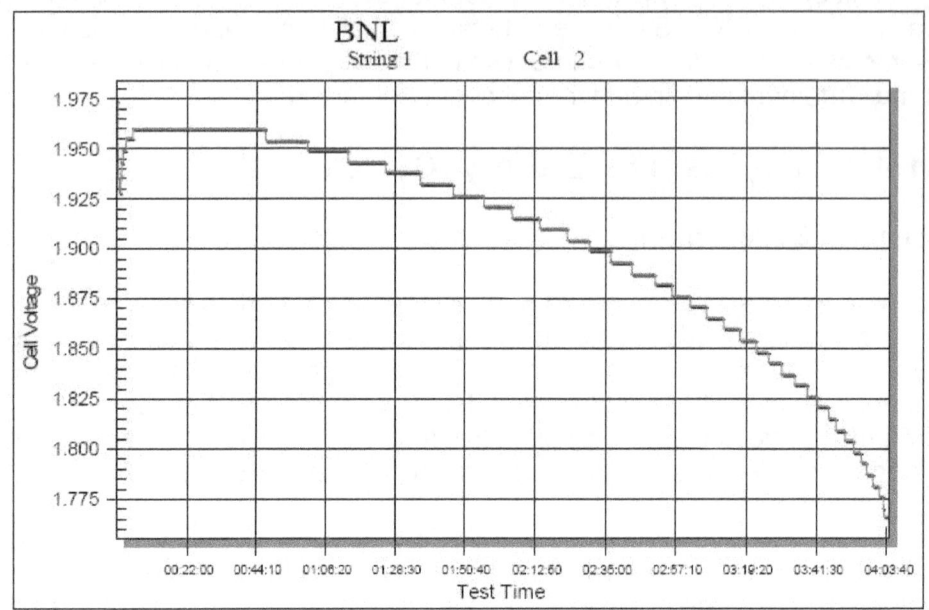

Figure 3-2   Enersys- cycle 1- cell 2 voltage response during discharge

Enersys Cycle 1 Recharge Voltage Profile: Once the load is removed from the battery string, the overall voltage increases to about 24.0 volts. The battery charger was reconnected and recharged at a float voltage of 27.0 volts for all 10 cycles of the Enersys testing as per the manufacturer's recommendations. Figure 3-3 illustrates the steady increase in voltage that occurs during recharge, reaching the terminal float voltage of 27.0 volts in about 11 hours for Enersys cycle 1.

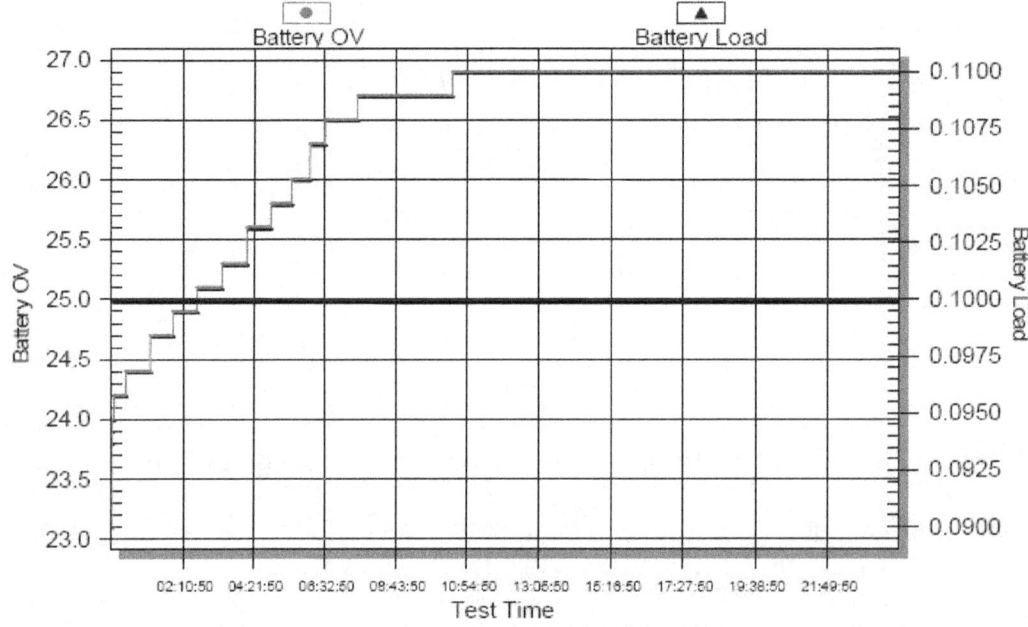

Figure 3-3   Enersys cycle 1 recharge voltage profile

16

The overall battery voltage response for the 10 Enersys test cycles is summarized in Table 3-1.

Table 3-1   Summary of Enersys battery string voltage response.

Cycle #	Starting/Recharge Voltage	Initial Voltage Drop Value	Time to Reach Recharge Voltage (Hours)
1	27	23.0	10.5
2	27	23.2	8.5
3	27	23.2	8.5
4	27	23.3	8.7
5	27	23.3	16.5
6	27	23.3	7.0
7	27	23.3	17.2
8	27	23.3	10.7
9	27	23.4	15.0
10	27	23.4	6.0

3.1.2.   GNB Battery String

Cycles 1 and 2 of the 10 test cycles of the GNB battery string were conducted at a recharge voltage of 28.0 volts. The remaining cycles were conducted at 27.0 volts. This vendor permits recharge to be conducted at the equalizing voltage which is the approach taken by some nuclear power plants.

GNB Cycle 1 Discharge Voltage Profile (28.0 volts): As illustrated in Figure 3-4, when the load is applied to the battery, the overall voltage decreases to about 23.2 volts. In this first cycle, the battery string voltage reached the automatic shutdown point of 21.0 volts in nearly four hours and six minutes which translates to a battery capacity of 102.4%.

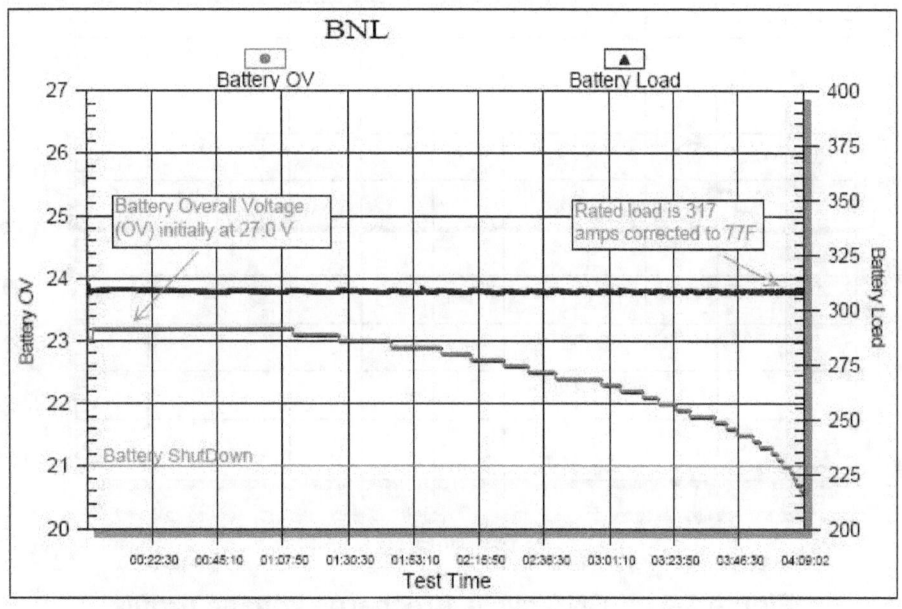

Figure 3-4   GNB cycle 1 discharge voltage profile

17

Figure 3-5 illustrates an individual cell's voltage response during the discharge iteration. Note the initial dip in voltage when the load is applied and the recovery as a stable current is established (Coup de Fouet effect).

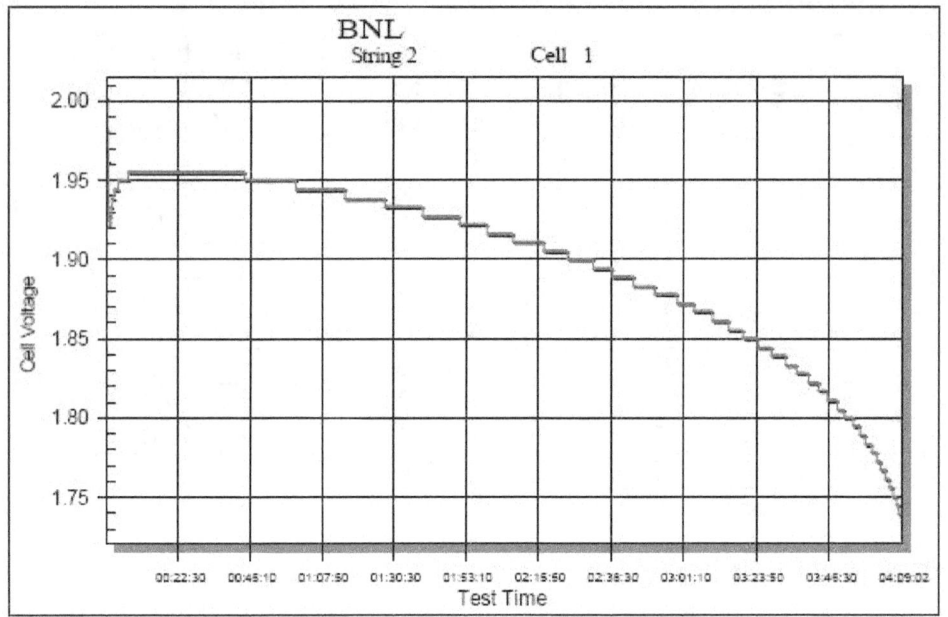

Figure 3-5   GNB cycle 1- cell 1 voltage response during discharge

GNB Cycle 1 Recharge Voltage Profile (28.0 volts): During recharge at an equalizing voltage of 28.0 volts, the battery charger stays in current limit mode longer and the battery string reaches the charger terminal voltage in approximately 8.5 hours. This is illustrated in Figure 3-6.

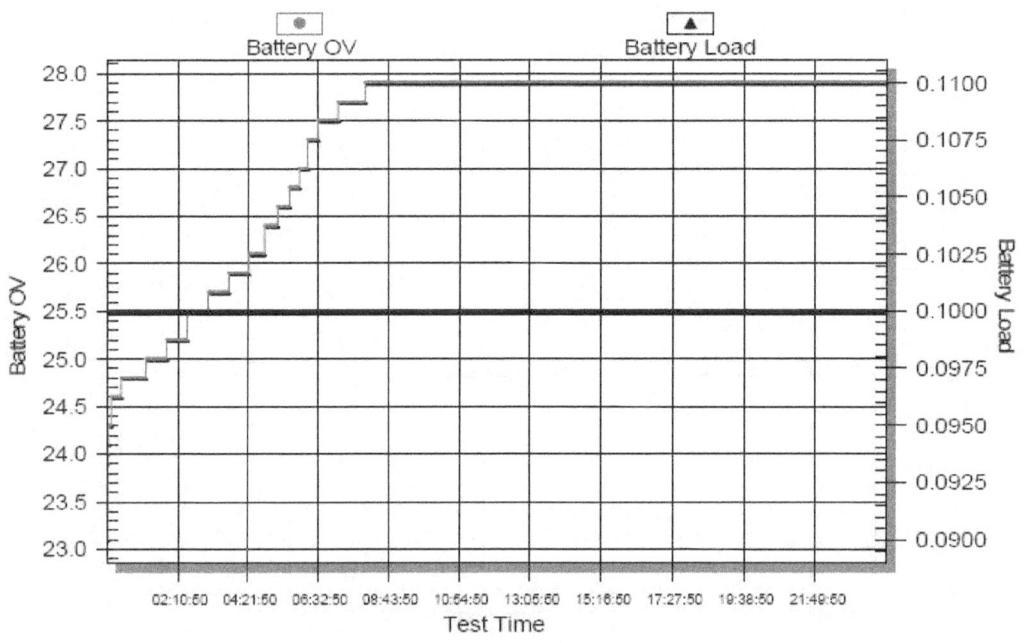

Figure 3-6   GNB cycle 1 recharge voltage profile

GNB Cycle 4 Discharge Voltage Profile (27.0 volts): In this cycle (Figure 3-7), the battery string voltage response to the automatic shutdown voltage of 21.0 volts does not vary significantly from the discharge response with the battery string starting at 28.0 volts.

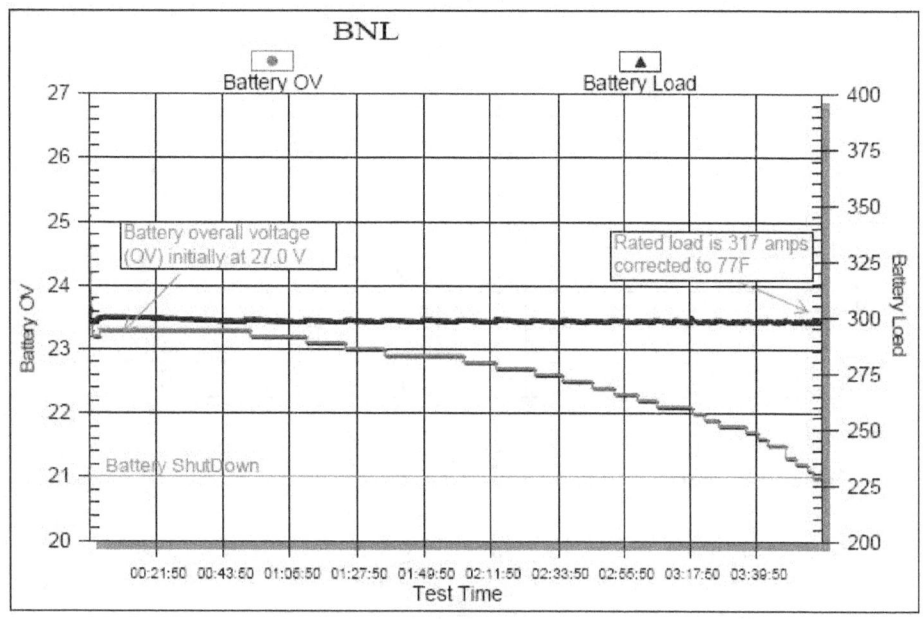

Figure 3-7   GNB cycle 4 discharge voltage profile

GNB Cycle 3 Recharge Voltage Profile (27.0 volts): As illustrated in Figure 3-8, recharging at 27.0 volts resulted in a significantly longer time to achieve the battery rated voltage (~13.5 hours vs. 8.5 hours that is illustrated in Figure 3-6). However, in later test cycles where the recharge was conducted at 27.0 volts, the battery string reached its terminal voltage of 27 volts within 8.0 hours. Table 3-2 summarizes the overall battery voltage response for the 10 cycles of testing of the GNB battery.

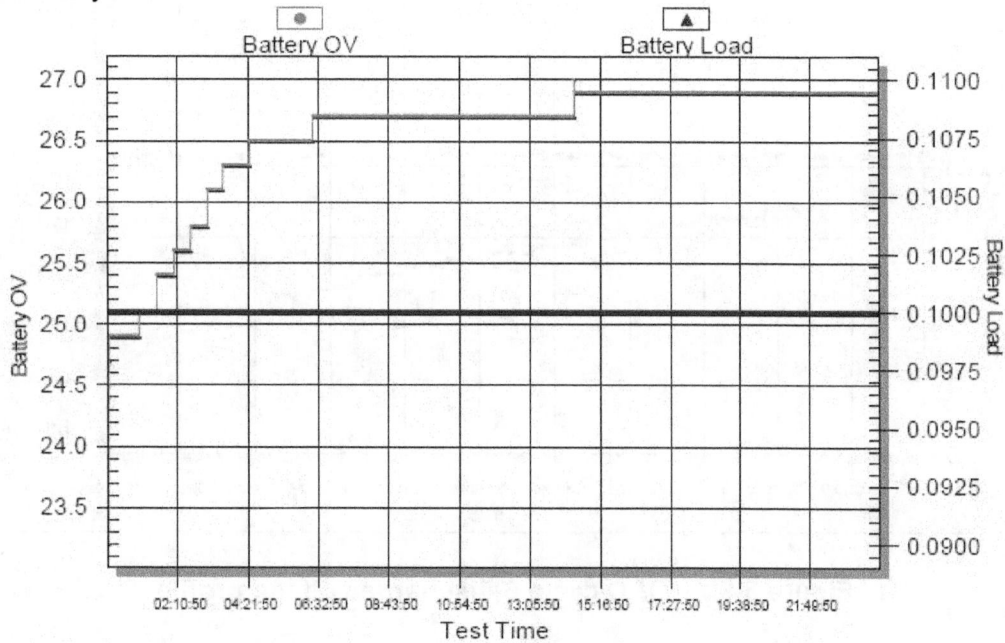

Figure 3-8   GNB cycle 3 recharge voltage profile

19

Table 3-2   Summary of GNB battery string voltage response.

Cycle #	Starting/Recharge Voltage	Initial Voltage Drop Value	Time to Reach Recharge Voltage (Hours)
1	28	23.0	8.4
2	28	23.1	8.1
3	27	23.1	13.5
4	27	23.2	10.5
5	27	23.2	8.0
6	27	23.2	12.5
7	27	23.2	6.6
8	27	23.2	8.5
9	27	23.3	6.5
10	27	23.3	8.7

3.1.3.   C&D Battery String

As with the GNB battery string, the C&D cycle testing was conducted at both a float voltage level of 27.0 volts and an equalizing voltage level of 28.0 volts. In cycle 6, conducted at 27.0 volts, Figure 3-9 illustrates the steady voltage decrease that occurred towards the automatic shutoff voltage of 21.0 volts. Figure 3-10 depicts the recharge voltage profile. Note that it takes approximately 17 hours for the battery string to reach its required terminal voltage of 27.0 volts. Table 3-3 summarizes the battery string voltage response for the 10 C&D cycles.

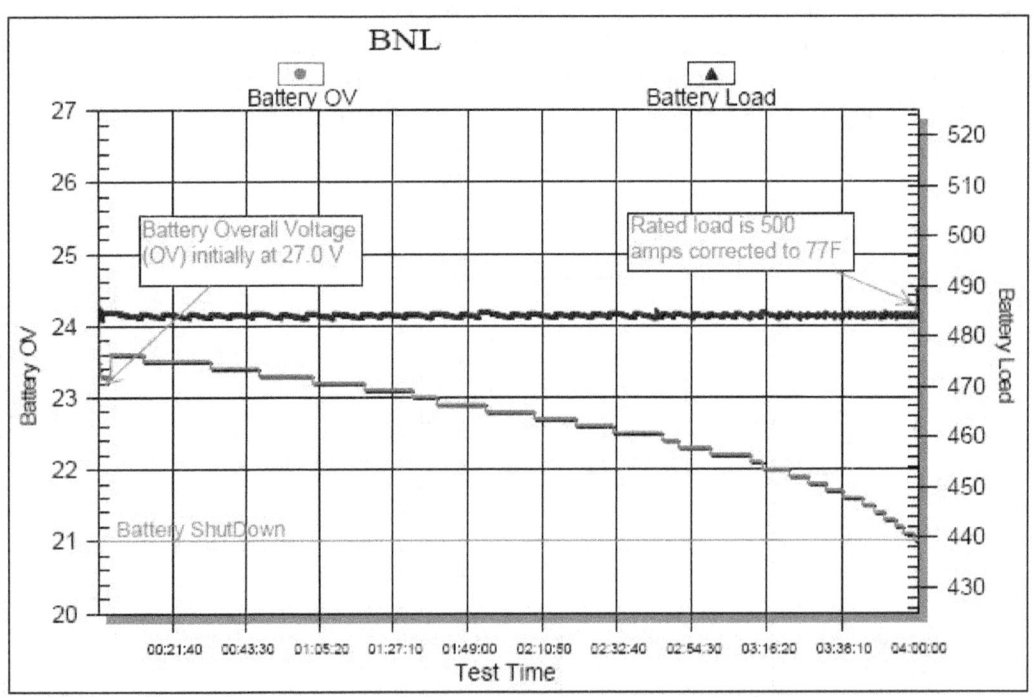

Figure 3-9   C&D cycle 6 discharge voltage profile

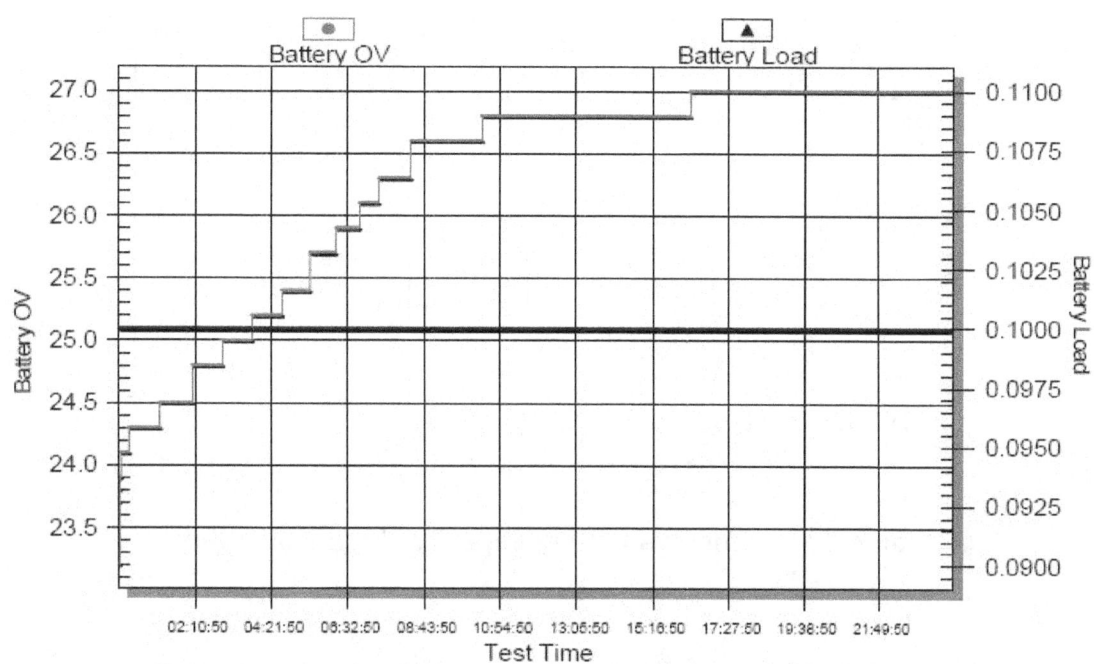

Figure 3-10   C&D cycle 6 recharge voltage profile

Table 3-3   Summary of C&D battery string voltage response.

Cycle #	Starting/Recharge Voltage	Initial Voltage Drop Value	Time to Reach Recharge Voltage (Hours)
1	27	23.1	14.25
2	27	23.2	12.25
3	28	23.2	14.75
4	28	23.3	11.25
5	27	23.3	18.5
6	27	23.3	16.5
7	27	23.3	11.0
8	27	23.4	10.5
9	28	23.3	15.5
10	28	23.4	11.0

3.1.4.   Battery Capacity Performance Tests

The battery capacity test set compares the time that it takes for the battery to reach 21.0 volts (1.75 volts/cell) to the rated (100%) performance time as the measure of battery capacity. This was calculated for every test cycle that was conducted. A trend of the capacity values for the 10 cycles conducted on each of the battery strings is portrayed in Figures 3-11, 3-12, and 3-13. Note that in all cases, the battery achieved greater than 100% capacity performance in the first cycle, with the capacity decreasing by 5-10% by the 10th cycle. One of the battery vendors stated that the battery should be able to be cycled 50-100 times before reaching an end of life state. While the cycling was not intended to age the battery, it is indicative of the number of performance tests that would be experienced by a typical battery over its 20-year life.

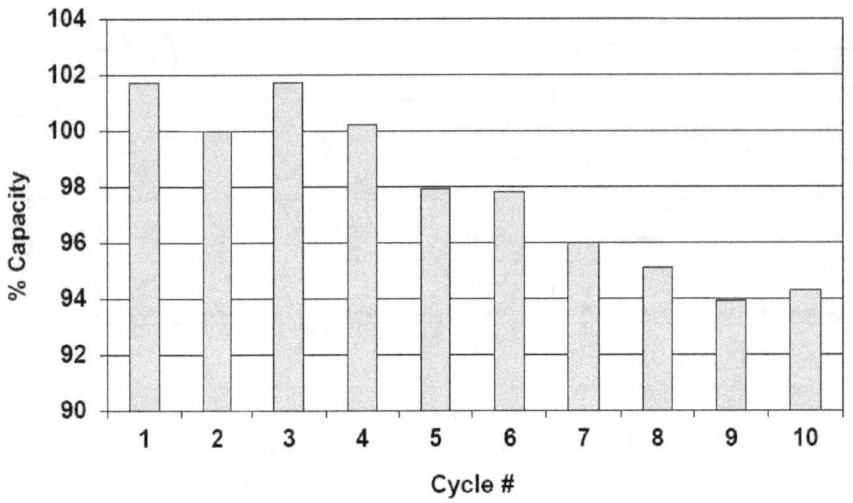

Figure 3-11   Enersys battery capacity change over time

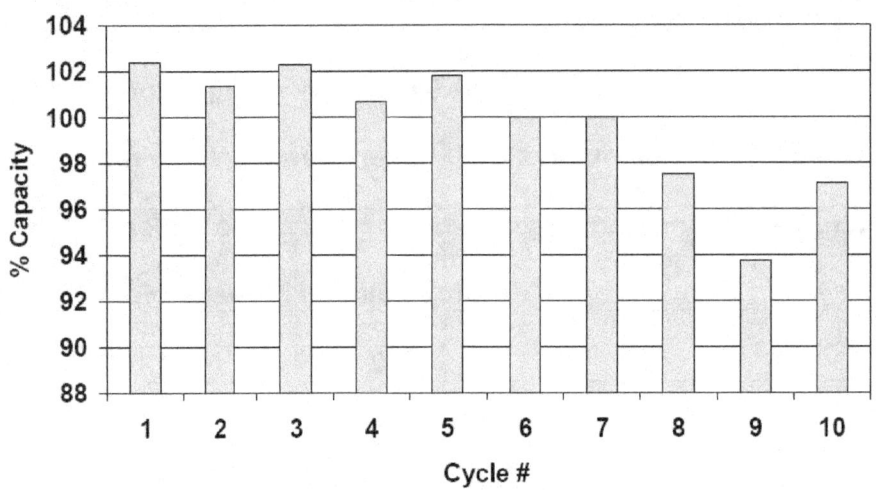

Figure 3-12   GNB battery capacity change over time

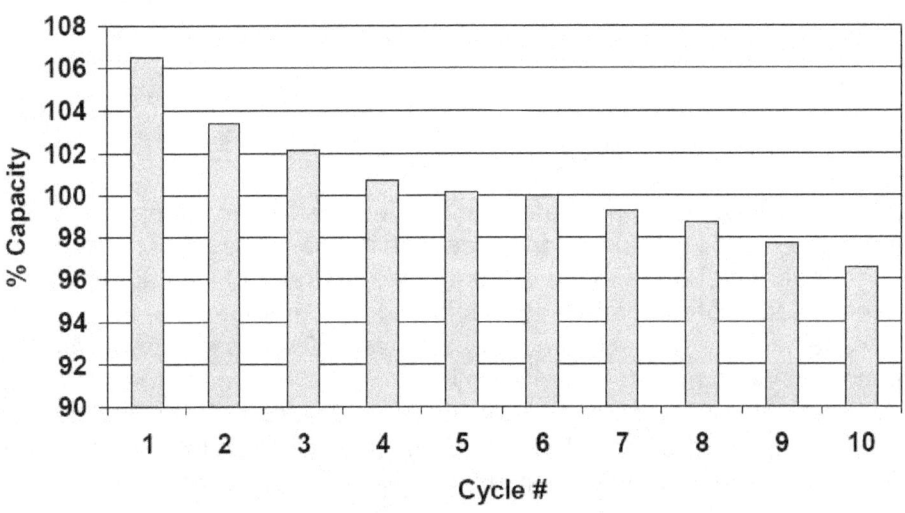

Figure 3-13   C&D battery capacity change over time

22

3.2. Float Current Responses on Recharge

Monitoring float current response during the recharge of a battery following a performance test was a key objective of this test program. Our initial testing set up for the first battery employed a float current monitoring device manufactured by Polytronics. For the second and third battery strings, we used a calibrated shunt as the primary current data and the Polytronics unit as a backup. An offset was detected in the Polytronics unit that required correction of the Enersys float current data. The data included in this section reflect the correction.

As illustrated in this section, the response varies somewhat from cycle to cycle and certainly between battery types. What is common is the general shape of the float current response curve and the approximate time at which a stable float current is reached, although the latter is sensitive to the voltage applied during recharge. More detailed analysis of the data represented by these curves and tables is provided in Section 4. This section contains observations of the float current curve characteristics as measured by the calculated curve time constant as well as the calculation of amp-hours returned to the battery.

Enersys Float Current Response

In Figure 3-14, the float current is shown to be constant at the current limit setting of the battery charger (180 amps) for approximately six hours at which time it starts decreasing in an exponential-like manner. The latter part of the curve in Figure 3-14 is expanded in Figure 3-15 to show that at about 24-hours a stable float current is achieved at about 0.5 amps. Slight variations were observed for the Enersys battery over the 10 test cycles; these data are summarized in Table 3-4. Note that all recharge cycles for the Enersys battery string were performed at the high end of the specified float voltage (27.0 volts) as recommended by the vendor.

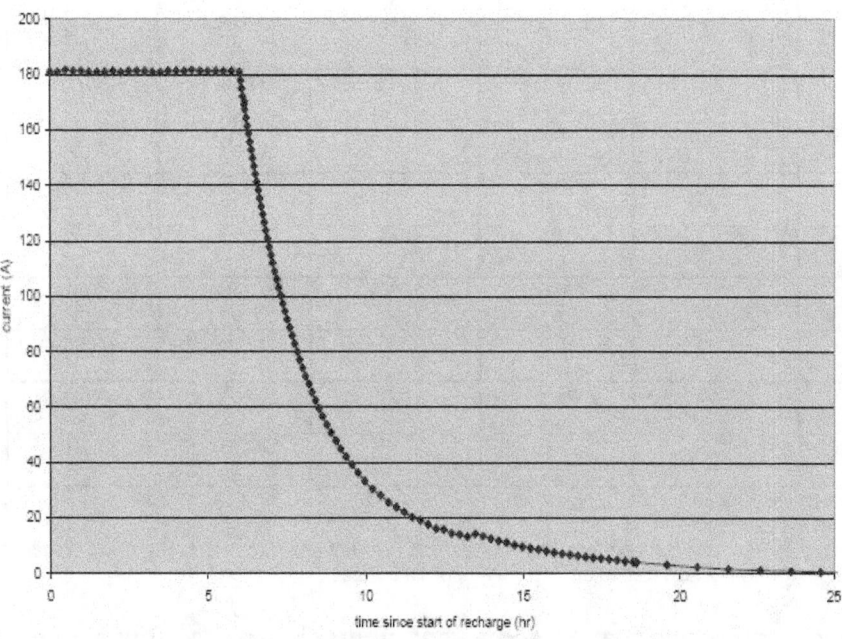

Figure 3-14   Enersys cycle 5 recharge current

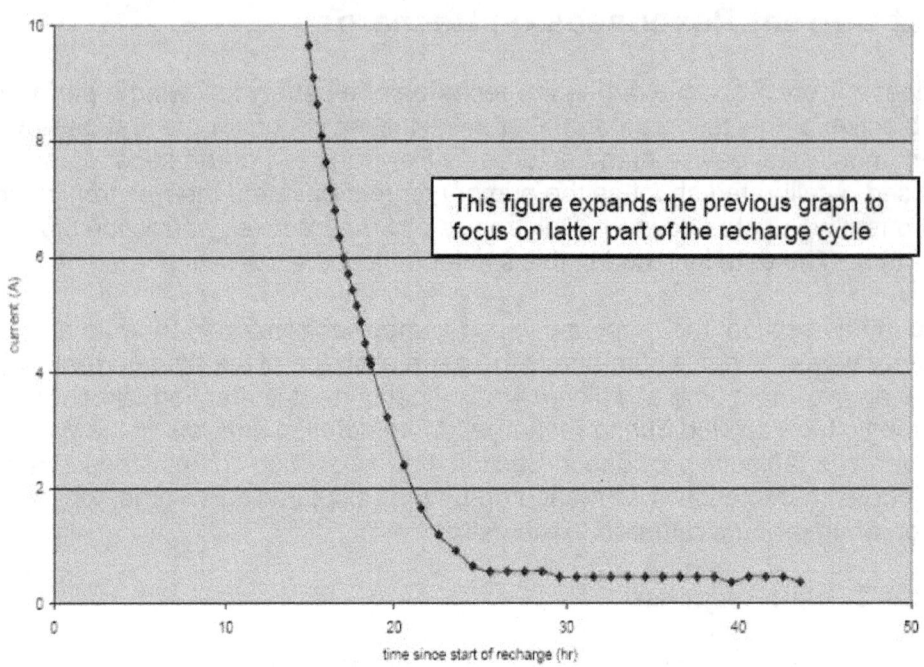

Figure 3-15   Enersys cycle 5 float current (late recharge)

Table 3-4   Summary of Enersys recharge/float current characteristics.

Cycle Number	Ah Discharged	Time to 100% Recharge (hr)	Time to 2 A (hr)	% Recharged at 2 A	Time Constant t_C (hr)	Time at Current Limit, Δt_{CL} (hr)	$\Delta t_{CL} + 5^*t_C$	% Recharged at $\Delta t_{CL} + 5^*t_C$
1 (73 F)	1530 Ah	11.2	11.9	101%	1.98	6.50	16.4	101% (1 amp)
2 (70 F)	1480 Ah	11.9	13.6	101%	2.27	6.00	17.4	102% (1 amp)
3 (68 F)	1486 Ah	13.8	14.8	101%	2.34	5.76	17.5	102% (3 amp)
4 (69 F)	1468 Ah	13.6	19.9	103%	2.38	5.66	17.6	102% (4 amp)
5 (74 F)	1516 Ah	14.0	20.6	103%	2.26	6.06	17.4	102% (5 amp)
6 (73 F)	1466 Ah	13.7	22.2	104%	2.18	5.61	16.5	102% (7 amp)
7 (73 F)	1496 Ah	15.8	26.0	103%	2.24	5.51	16.7	100% (9 amp)
8 (74 F)	1440 Ah	15.8	26.3	104%	2.16	5.28	16.1	100% (9 amp)
9 (73 F)	1429 Ah	18.3	26.8	102%	2.17	5.24	16.1	99% (9 amp)
10 (70 F)	1395 Ah	17.2	28.6	104%	2.16	4.87	15.7	99% (10 amp)

Rated discharge capacity at 385 amps for four hours at 77 F is 1540 Ah.
Recharge for all cycles was conducted at 27 volts.
Columns 3, 5, and 9 are based upon comparisons of the measured Ah discharged (col. 2) to the rated Ah capacity for a 4-hr discharge performance test from 2.25 V/cell to an end voltage of 1.75 V/cell.

In Table 3-4, it is worth noting that we somewhat arbitrarily selected 2 amps as a point of comparison among the three battery types. The ampere-hours returned to the battery was calculated from the float current response curve and compared to the actual amount of ampere-hours that were discharged as determined by the capacity test set data. Finally, the time

24

constant of the float current response curve was calculated using a process described in more detail in Appendix A.

GNB Float Current Response

The Exide GNB nuclear grade batteries, provided by Nuclear Logistics, Inc., were recharged at an equalizing voltage of 28.0 volts for cycles 1 and 2 and at the float voltage level of 27.0 volts for the remaining eight cycles in order to assess the difference in the float current response. Feedback obtained from the battery vendors indicated that some of their customers (the nuclear power plants) used an equalizing charge following a discharge test to speed up the electrolyte mixing process. We therefore wanted to determine how the float current responses would differ under these two recharge conditions. This section illustrates the float current response for representative cycles of recharge at a float voltage of 27.0 volts and an equalizing charge of 28.0 volts. All 10 cycles of recharge data are then summarized in Table 3-5.

Figure 3-16 illustrates the recharge/float current response with the battery charger set at its equalizing voltage of 28.0 volts. The battery stays in a current limit mode (180 amps in this case) for more than six hours. Once the battery approaches 28.0 volts, the current from the charger decreases rapidly towards an asymptotic float current value of less than 0.5 amps, as shown in Figure 3-17.

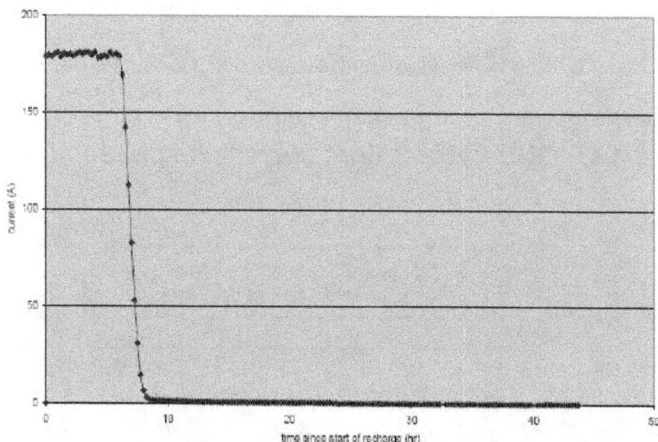

Figure 3-16   GNB cycle 1 float current response at 28.0 volts

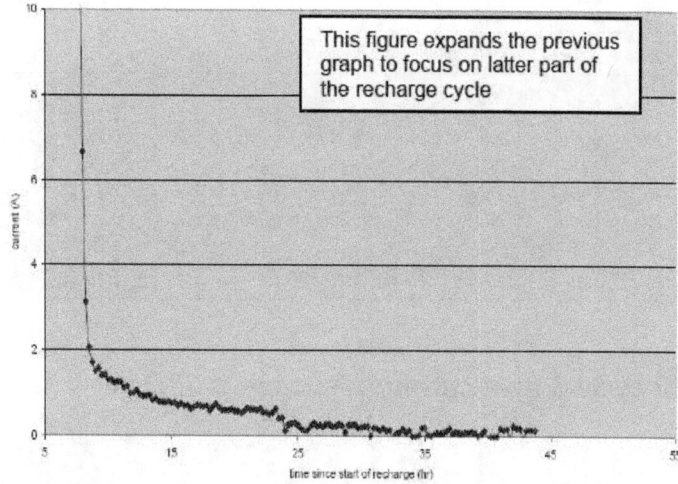

This figure expands the previous graph to focus on latter part of the recharge cycle

Figure 3-17   GNB Cycle 1 float current response at 28.0 volts (late recharge cycle)

25

Figures 3-18 and 3-19 depict the recharge/float current response with the battery charger set at 27.0 volts. The current limit mode is maintained for a shorter time (less than four hours) but the recharge/float current response is more gradual than the response at 28.0 volts. In both cases, the same stable float current is ultimately obtained.

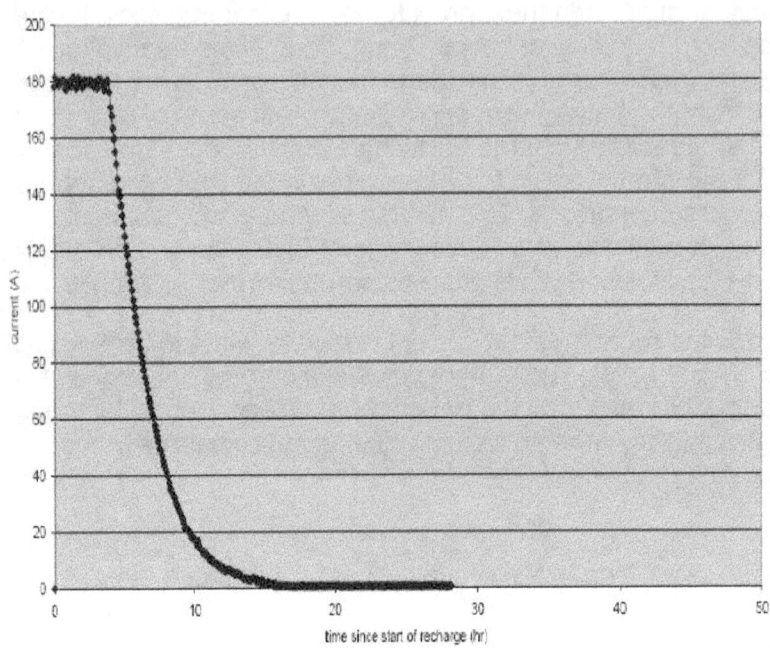

Figure 3-18   GNB cycle 3 float current response at 27.0 volts

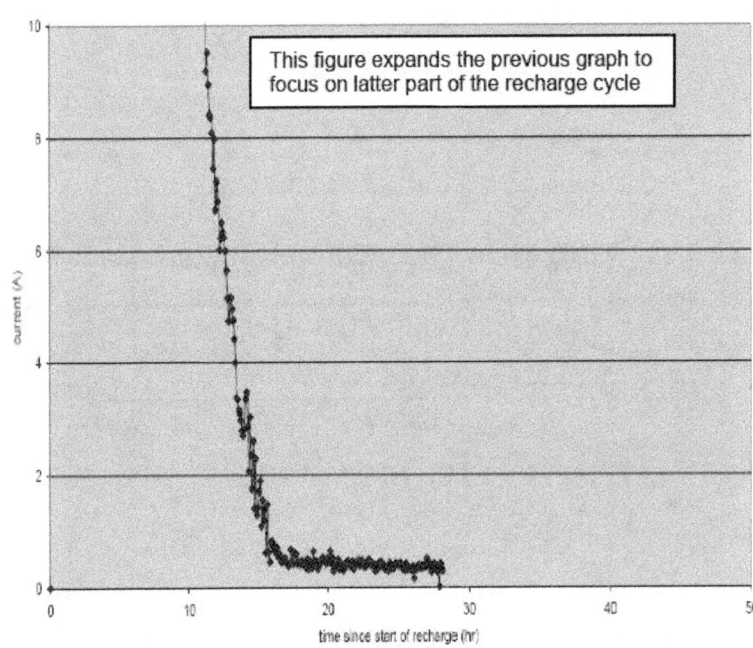

This figure expands the previous graph to focus on latter part of the recharge cycle

Figure 3-19   GNB cycle 3 float current response at 27.0 volts (late recharge cycle)

It was apparent from performing the recharge at an equalizing charge (28.0 volts) vs. a float charge (27.0 volts) that the characteristic curve was significantly different as follows:

1. The time the charger stays in current limit is longer for an equalizing charge
2. The time constant is shorter for an equalizing charge
3. The steady-state float current is reached sooner for an equalizing charge

The 10 cycles of the GNB float current data are summarized in Table 3-5:

Table 3-5   Summary of GNB recharge/float current characteristics.

Cycle Number	Ah Discharged	Time to 100% Recharge (hr)	Time to 2 A (hr)	% Recharged at 2 A	Time Constant t_c (hr)	Time at Current Limit, Δt_{CL} (hr)	$\Delta t_{CL} + 5*t_c$	% Recharged at $\Delta t_{CL} + 5*t_c$
1 (72 F)	1278	8.5	8.5	100%	0.97	6.34	11.2	100% (1 amp)
2 (65 F)	1191	7.3	8.8	102%	1.22	5.67	11.8	102% (1 amp)
3 (58 F)	1157	10.3	14.8	103%	2.90	3.92	18.4	103% (<1 amp)
4 (67 F)	1201	12.0	15.8	101%	2.51	4.25	16.8	101% (1 amp)
5 (64 F)	1190	13.3	16.6	101%	2.51	4.15	16.7	101% (2 amp)
6 (66 F)	1184	Detailed data are not available due to an interruption of power to the data acquisition systems.						
7 (64 F)	1168	15.0	18.8	101%	2.49	3.93	16.4	100% (4 amp)
8 (69 F)	1178	14.6	19.7	102%	2.30	4.15	15.7	100% (4 amp)
9 (71 F)	1151	15.8	17.9	101%	2.28	4.03	15.4	100% (4 amp)
10 (70 F)	1184	14.9	19.8	101%	2.24	4.25	15.5	100% (4 amp)

Rated discharge capacity at 317 amps for four hours at 77 F is 1268 Ah.
Recharge for Cycles 1 and 2 were conducted at 28.0 volts or 2.33 volts per cell (highlighted data).
Columns 3, 5, and 9 are based upon comparisons of the measured Ah discharged (col. 2) to the rated Ah capacity for a 4-hr discharge performance test from 2.25 V/cell to an end voltage of 1.75 V/cell.

As noted above, detailed data are not available for GNB cycle 6 due to an interruption of power to the data acquisition systems. However, the four-hour discharge at 296 amps and the subsequent recharge to 27 volts for seven days was unaffected by the electrical problem.

C&D Battery Float Current Response

The C&D LCR-33 battery is the largest in capacity of the three battery types tested. As a result, the recharge times were longer than the other two battery strings. Recharge of the C&D batteries occurred at both 27.0 volts and 28.0 volts to assess float current response in both cases. These voltages were chosen based on the vendor recommendations for float and equalize voltage values. The recharge for cycle 1 conducted at 27 volts is illustrated in Figures 3-20 and 3-21:

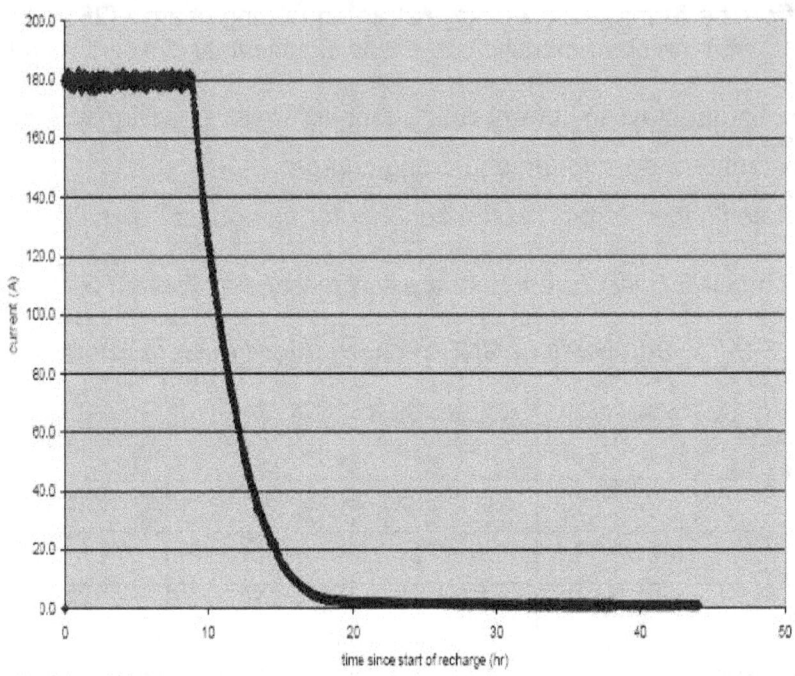

Figure 3-20   C&D cycle 1 charging/float current response at 27.0 volts

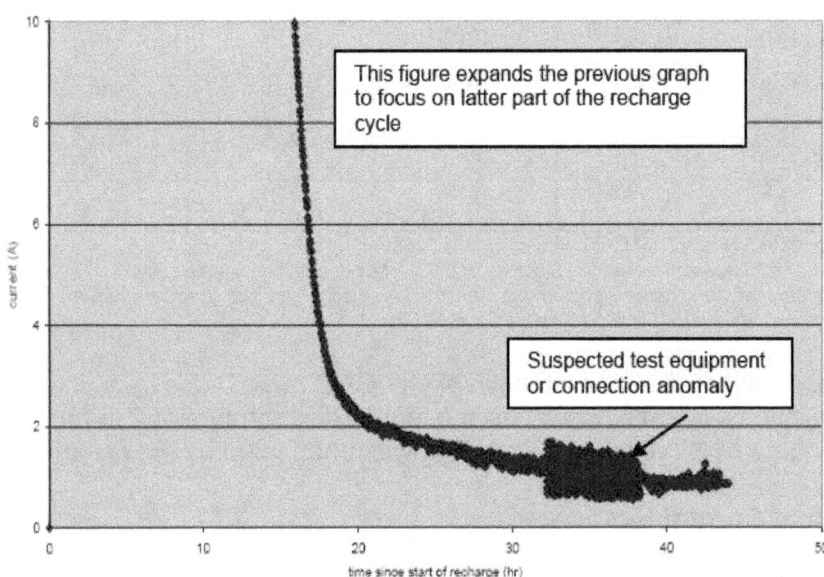

This figure expands the previous graph to focus on latter part of the recharge cycle

Suspected test equipment or connection anomaly

Figure 3-21   C&D cycle 1 float current response at 27.0 volts (late recharge cycle)

Figures 3-22 and 3-23 show the response with a recharge voltage of 28.0 volts up to 40 hours at which time the charger automatically switched to an output voltage of 27.0 volts. The float current dropped from about 2 amps to 0.5 amps when the charger output switched from 28.0 to 27.0 volts.

28

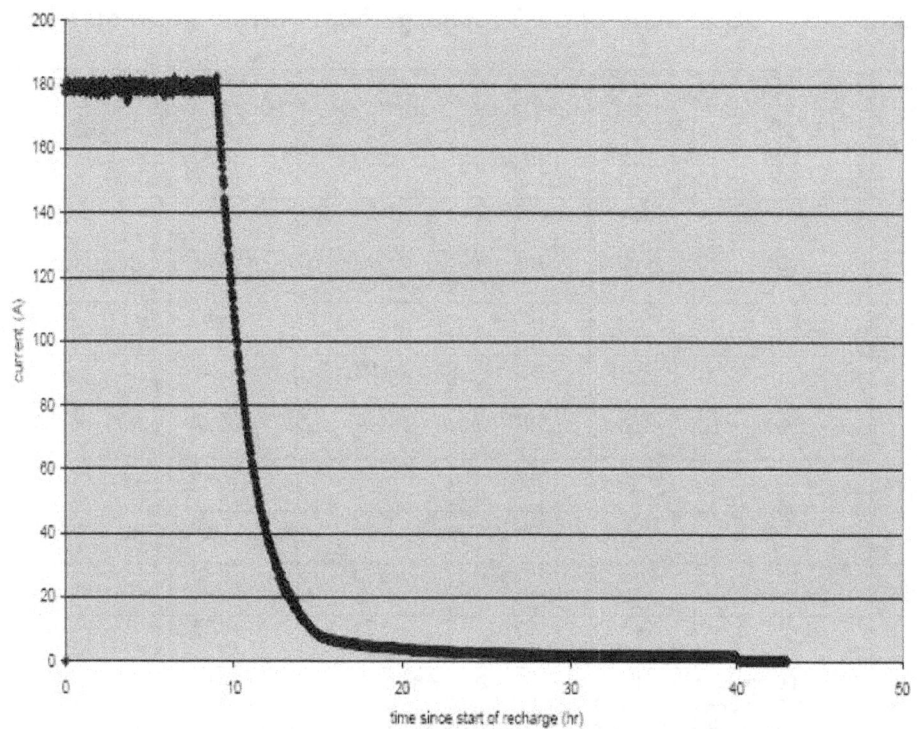

Figure 3-22   C&D cycle 10 recharge/float current response at 28.0 V

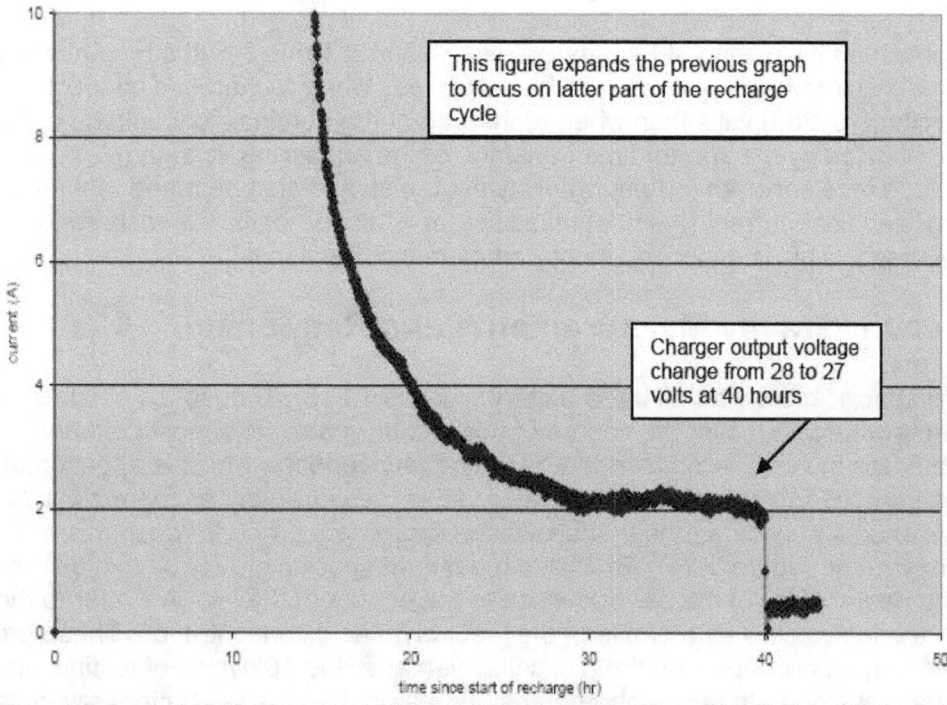

This figure expands the previous graph to focus on latter part of the recharge cycle

Charger output voltage change from 28 to 27 volts at 40 hours

Figure 3-23   C&D cycle 10 recharge/float current response at 28.0 V (late recharge cycle)

The recharge/float current data for all 10 cycles is summarized in Table 3-6. The highlighted rows indicate where recharge was conducted at 28.0 volts; otherwise 27.0 volts was used.

Table 3-6   Summary of C&D recharge/float current characteristics

Cycle Number	Ah Discharged	Time to 100% Recharge (hr)	Time to 2 A (hr)	% Recharged at 2 A	Time Constant t_c (hr)	Time at Current Limit, Δt_{CL} (hr)	$\Delta t_{CL} + 5^*t_c$	% Recharged at $\Delta t_{CL} + 5^*t_c$
1 (70 F)	2050	14.6	20.6	102%	3.03	8.87	24.0	102% (2 amp)
2 (72 F)	2007	14.8	21.3	102%	3.05	8.52	23.8	103% (2 amp)
3 (73 F)	2000	12.1	24.0	105%	1.78	9.85	18.8	104% (5 amp)
4 (72 F)	1956	11.9	24.0	105%	1.80	9.57	18.6	104% (4 amp)
5 (72 F)	1944 (97.2%)	14.9	25.5	105%	3.07	7.82	23.2	105% (5 amp)
6 (71 F)	1936 (96.8%)	18.3	28.6	103%	3.04	7.72	22.9	102% (5 amp)
7 (72 F)	1929 (96.5%)	18.7	30.2	103%	2.98	7.70	22.6	102% (5 amp)
8 (72 F)	1920 (96.0%)	19.4	31.3	103%	2.98	7.60	22.5	101% (6 amp)
9 (72 F)	1900	12.5	40.0	107%	2.00	8.90	18.9	104% (5 amp)
10 (72 F)	1901	12.1	38.7	107%	1.94	9.05	18.8	104% (4 amp)

Rated discharge capacity at 500 amps for four hours at 77 F is 2000 Ah.
Recharge for cycles 1-2 and 5-8 was conducted at 27 volts.
Recharge for cycles 3-4 was conducted at 28 volts for 24 hours, then reduced to 27 volts.
Recharge for cycles 9-10 was conducted at 28 volts for 40 hours, then reduced to 27 volts.
Recharge current data for cycle 5 are from the Polytronics unit.
Note: Highlighted data are associated with recharge at an equalizing voltage of 28.0 volts (2.33 volts per cell)

As can be observed by looking at Figures 3-20 to 3-23 and Table 3-6, the time needed to return 100% of the discharged amp-hours to the battery is less when recharge is conducted at the equalizing voltage of 28.0 volts than when recharged at the nominal float voltage of 27.0 volts. This is also reflected by the shorter time constant for the equalizing voltage (~2 hours) versus the float voltage time constant (~3 hours). However, it should also be noted that the time to reach a stabilized float current (2 amps) takes longer when the battery is recharged at an equalizing voltage. This is described more completely in Section 4.1.

3.3. Specific Gravity Measurements and Response

The nominal specific gravity for all three battery types is 1.215. Thousands of specific gravity measurements were taken over the course of this test program. The most common measurement was the one recommended by the battery vendors, which is approximately 1/3 of the distance from the top of the plates; we refer to this measurement as the midpoint reading. The battery vendors provide a tube that allows the instrument probe to reach this point on a consistent basis (see Figure 3-24). BNL also measured specific gravity on two cells on each battery string near the top of the cell and close to the bottom of the jar. We refer to the three readings on these two cells as a profile of the electrolyte acid concentration. These profile readings were repeated weekly on the two cells even after the 10 cycles of testing were completed for that string. In fact, eight months of data on the Enersys battery revealed that it takes a very long time for equilibrium to be reached following a series of performance tests. Weekly specific gravity profile testing on the GNB (seven months) and C&D (four months) battery strings also supported this observation. The impact of stratification on battery capacity and capability is measureable but not operationally significant as described further in Section 4.3.

Also provided in this section are data that clearly reveal the consistency among cells as the discharge and recharge conditions occur. These data support the use of pilot cells in a battery string as a reasonable method of determining the state of battery specific gravity for the string.

Specific Gravity
Sampling Tube

Figure 3-24   Specific gravity sampling tube

Enersys Battery String:

Measurements of specific gravity were taken during the discharge of the battery (four hour performance test) and regularly during the recharge of the battery. Figure 3-25 illustrates the specific gravity response over 25 hours for a typical discharge/recharge cycle on the Enersys battery. Figure 3-26 depicts the response of the specific gravity for a typical weekly test cycle. The last set of readings of the battery on one test cycle became the starting point for the next test cycle, with one performance test performed per week. Furthermore, even after the testing was completed on a battery string, specific gravity profile measurements continued to be taken weekly. For the Enersys battery string, which was the first battery tested, the readings continued for more than eight months while the battery remained on a float charge. It took that long for uniformity to be reached in the specific gravity measurements at the top, middle, and bottom of the cell. The eight months needed for the electrolyte in the Enersys battery to reach equilibrium is illustrated in Figure 3-27.

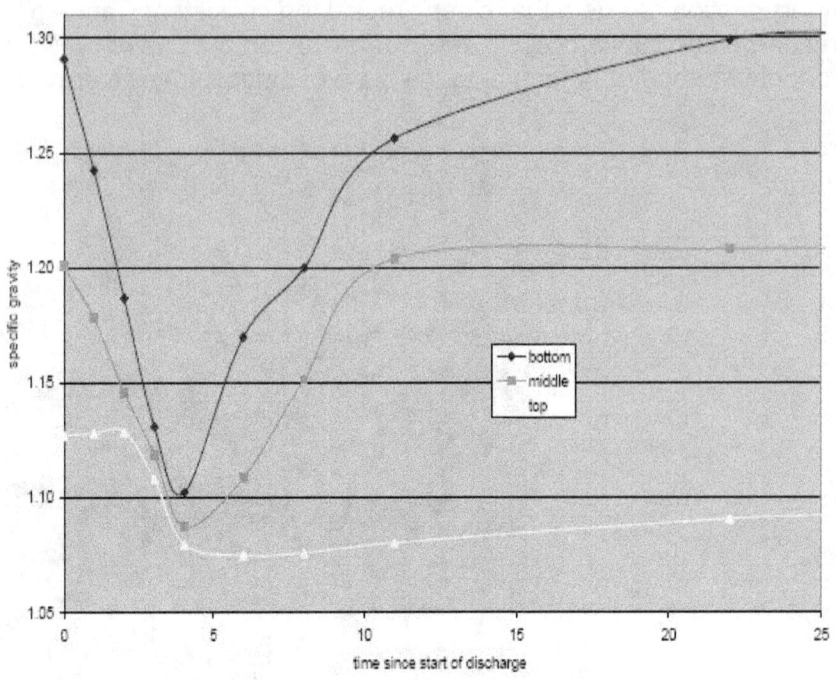

Figure 3-25   Enersys cycle 5 early specific gravity response

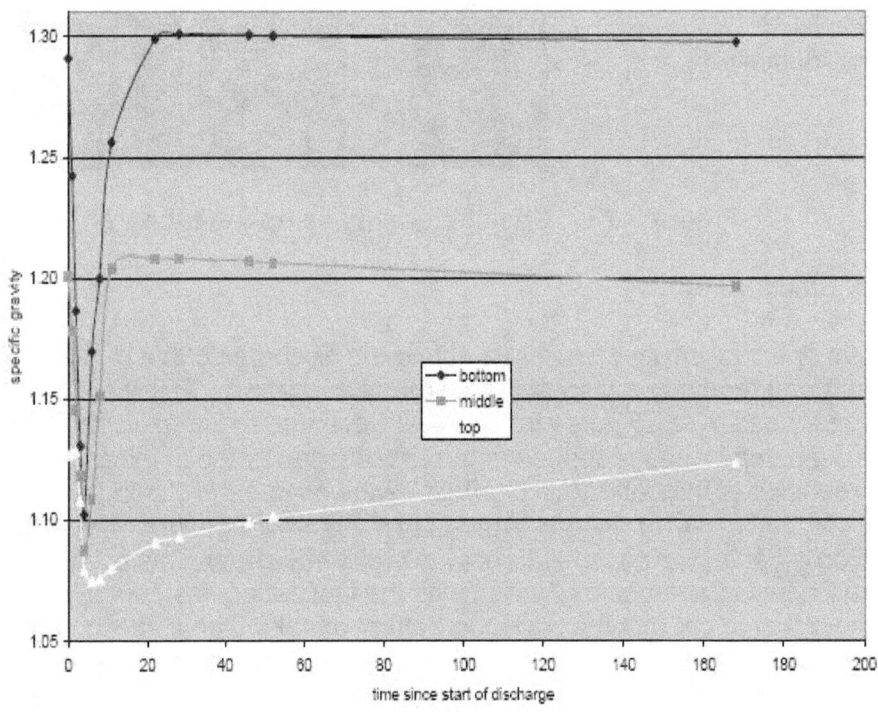

Figure 3-26   Enersys cycle 5 full test cycle specific gravity response

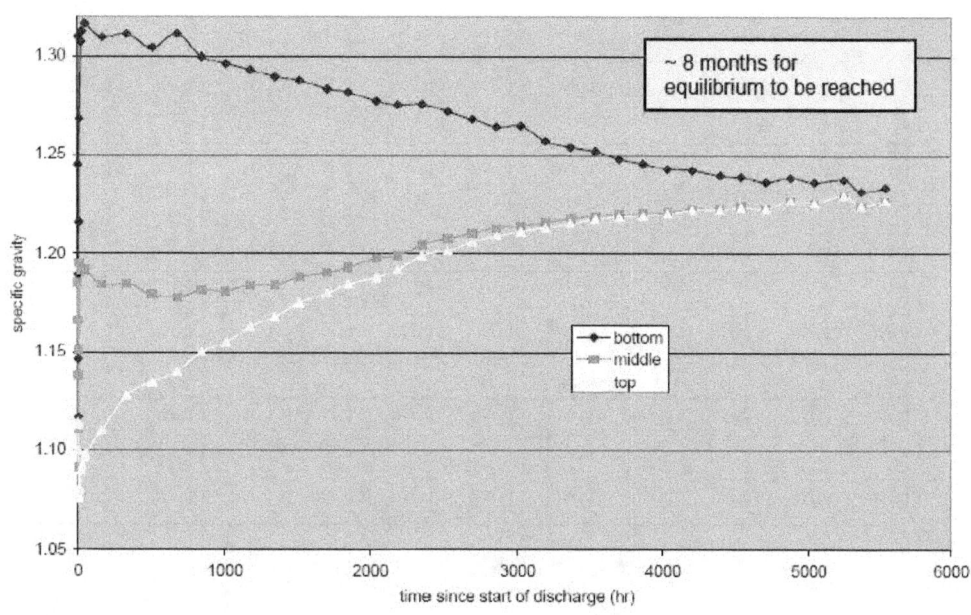

Figure 3-27   Enersys very long term specific gravity response (weekly readings)

GNB Battery String:

The GNB battery, supplied by Nuclear Logistics Inc., was the second battery string to be tested. As indicated in the following figures (Figures 3-28 to 3-31), the specific gravity response during the performance test and recharge is very similar to what was seen with the Enersys battery. Similarly, Figure 3-32 shows that this battery took about seven months on a float charge for the electrolyte to reach equilibrium. The GNB cycle 1 testing is used as an example and is compared with GNB cycle 3. For cycle 1 (Figures 3-28 and 3-29), an equalizing voltage of 28.0 volts was used during the recharge cycle. For cycle 3 (Figures 3-30 and 3-31), a float voltage of 27.0 volts was used during recharge.

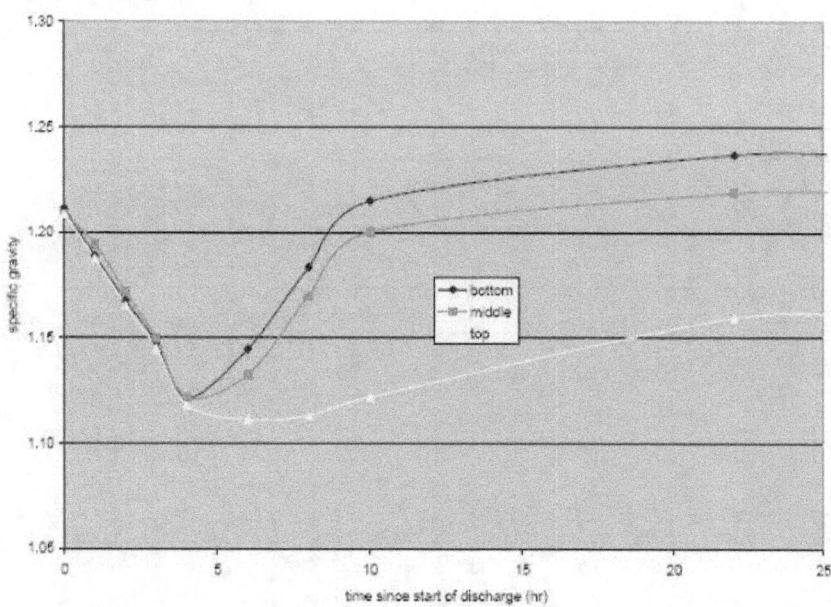

Figure 3-28   GNB cycle 1 early specific gravity response

33

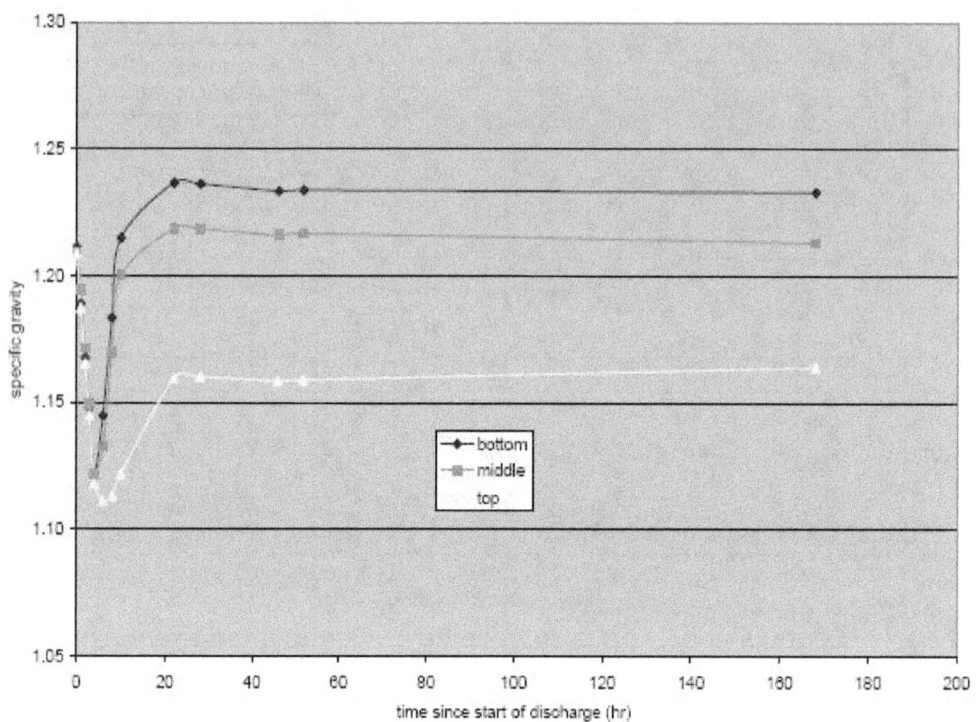

Figure 3-29   GNB cycle 1 specific gravity full test cycle response

In cycle 1, note that the midpoint specific gravity reading is restored to its initial value in about 15 hours and is stable within 24 hours. In cycle 3 (Figures 3-30 and 3-31), with a recharge voltage of 27.0 volts, it was observed that the specific gravity response was virtually identical.

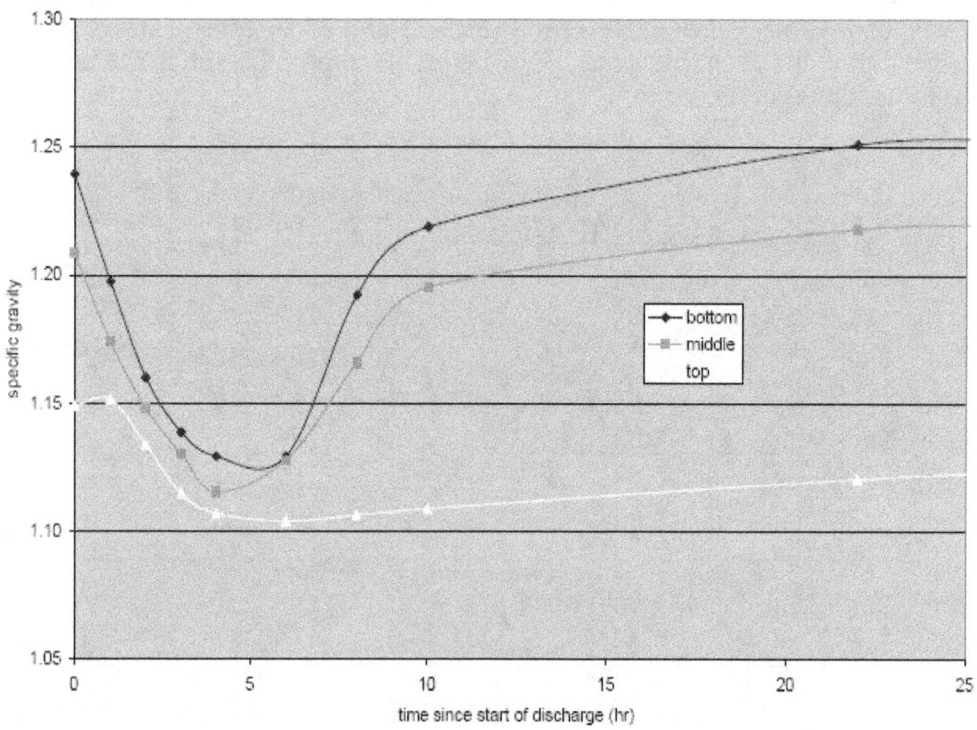

Figure 3-30   GNB cycle 3 specific gravity early cycle response

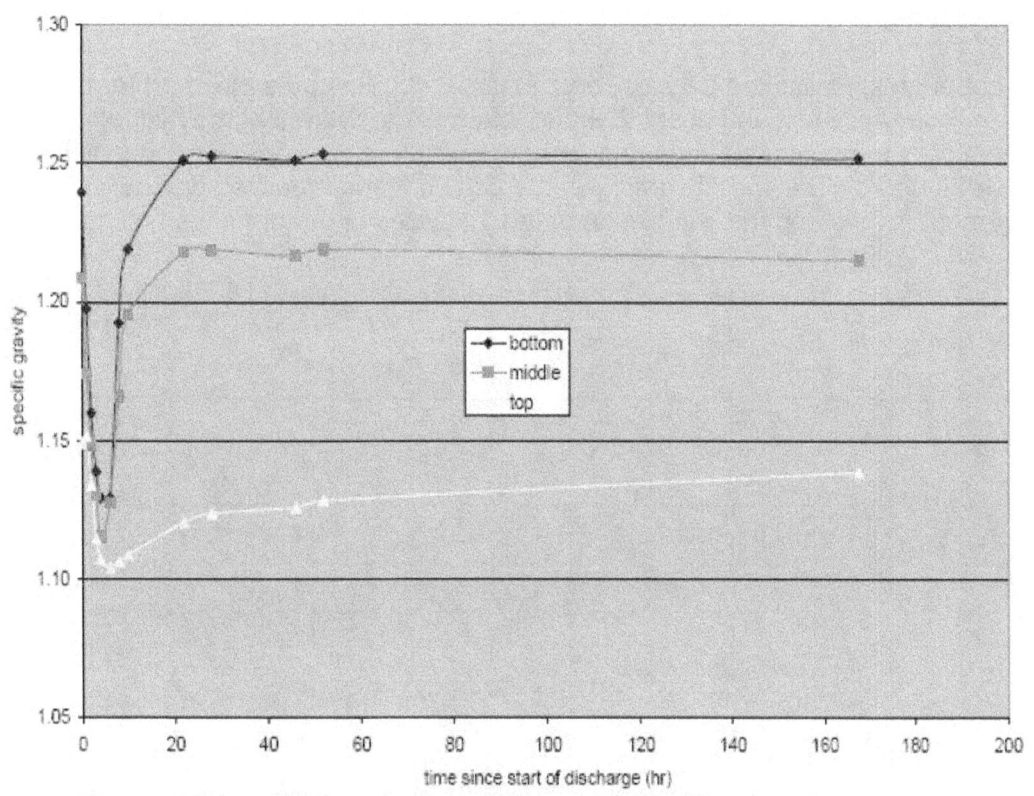

Figure 3-31   GNB cycle 3 specific gravity full test cycle response

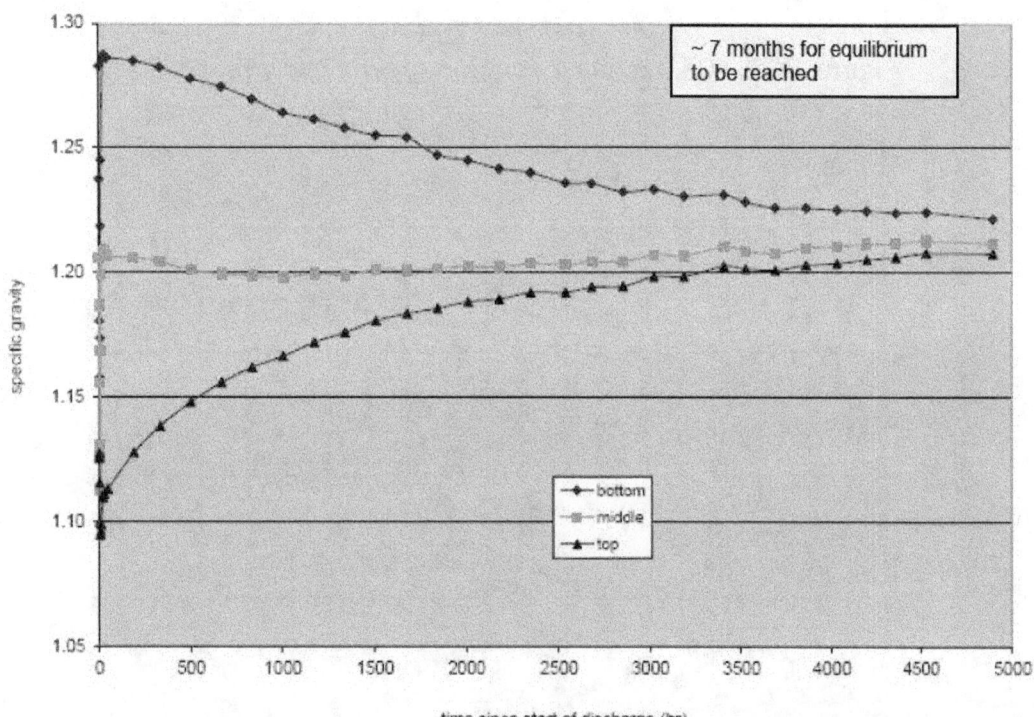

Figure 3-32   GNB specific gravity very long term response (weekly readings)

C&D Battery String:

The third battery tested was the C&D battery. Figure 3-33 illustrates the short term response during the performance test and about 20 hours afterwards. Note that the midpoint specific gravity is restored to its initial value during recharge within about 15 hours and is at its stable value within 24 hours. The specific gravity in the C&D batteries reached an equilibrium state within four months following the completion of the 10 cycles of performance testing as illustrated in Figure 3-35.

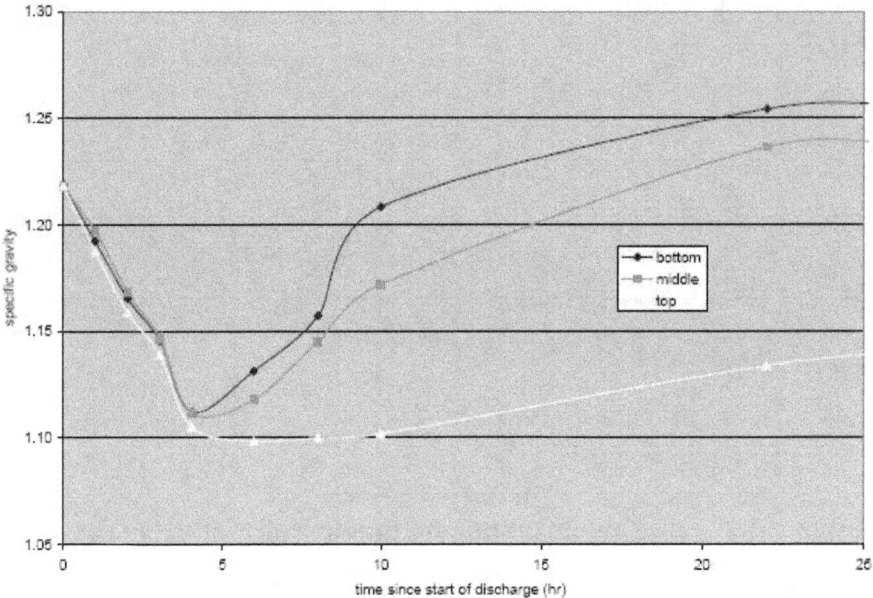

Figure 3-33   C&D cycle 1 specific gravity early response

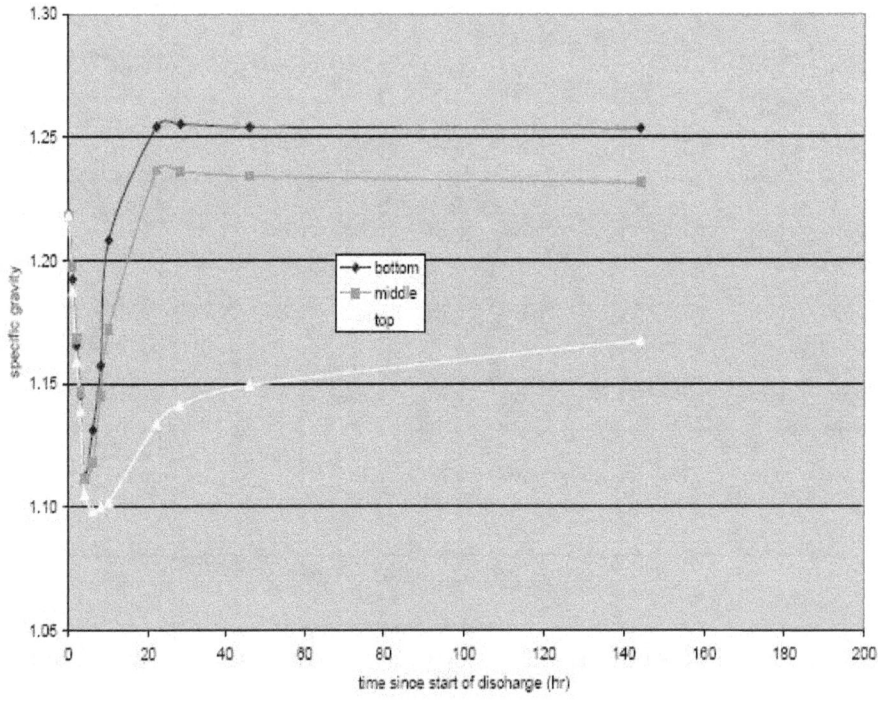

Figure 3-34   C&D cycle 1 specific gravity full test cycle response

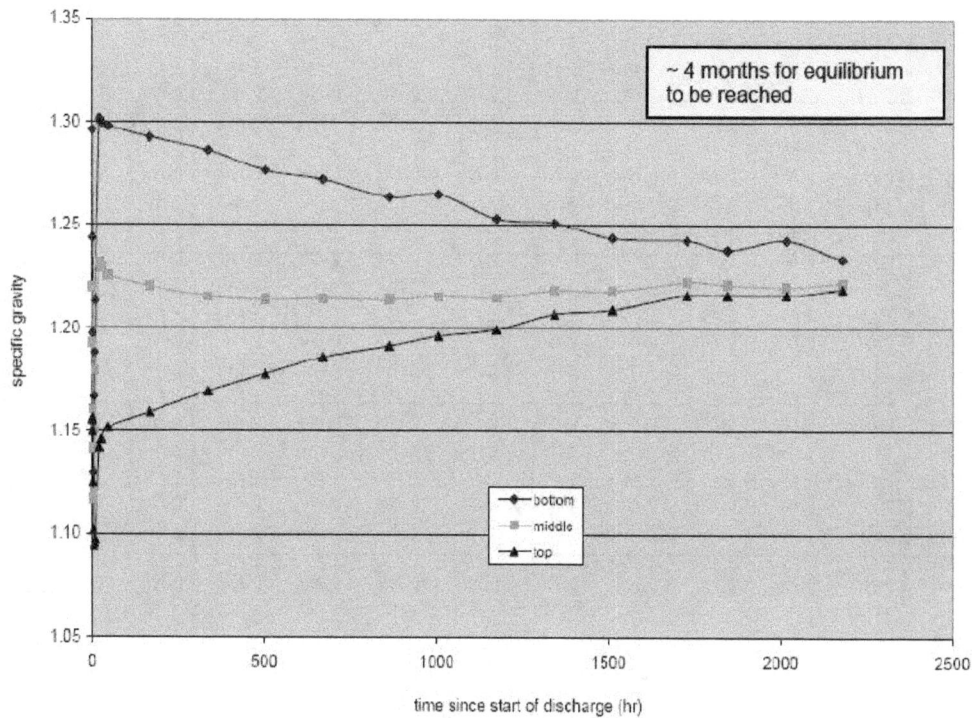

Figure 3-35   C&D specific gravity very long term response (weekly readings)

Another aspect of specific gravity is the response in each cell. This is important to know since nuclear power plants rely on using data from pilot cells to ascertain the condition for the entire battery string. Figures 3-36 to 3-38 provide examples of the cell midpoint specific gravity readings that indicate that the response among the cells is uniform and that the use of a pilot cell is appropriate.

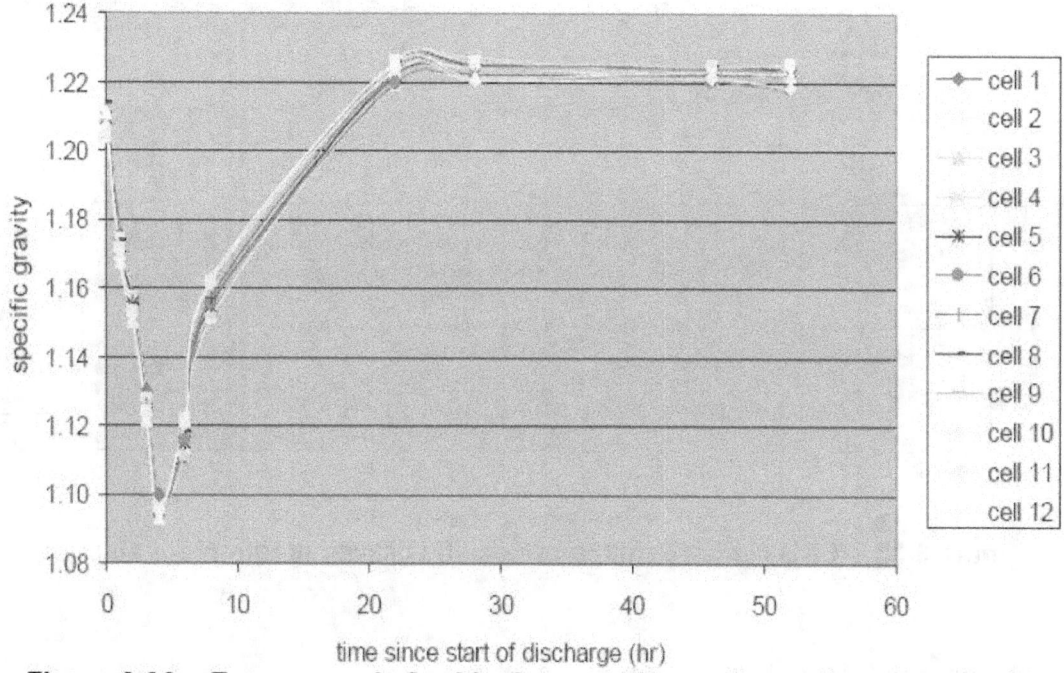

Figure 3-36   Enersys cycle 2 midpoint specific gravity readings for all cells

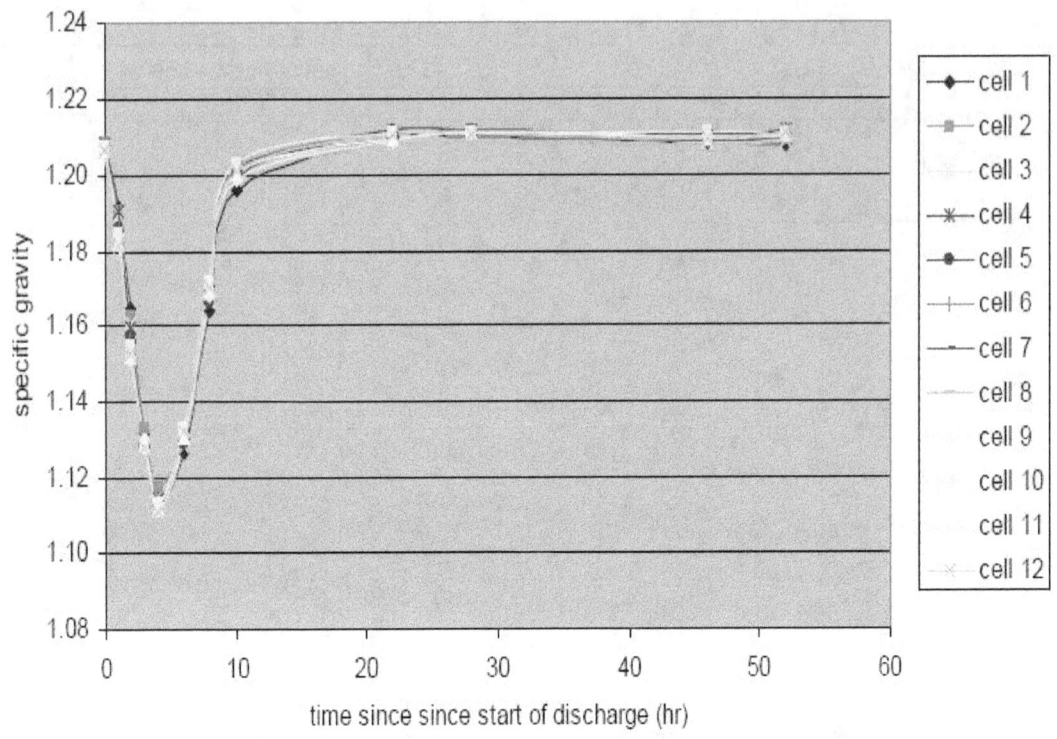

Figure 3-37   GNB cycle 8 midpoint specific gravity readings for all cells

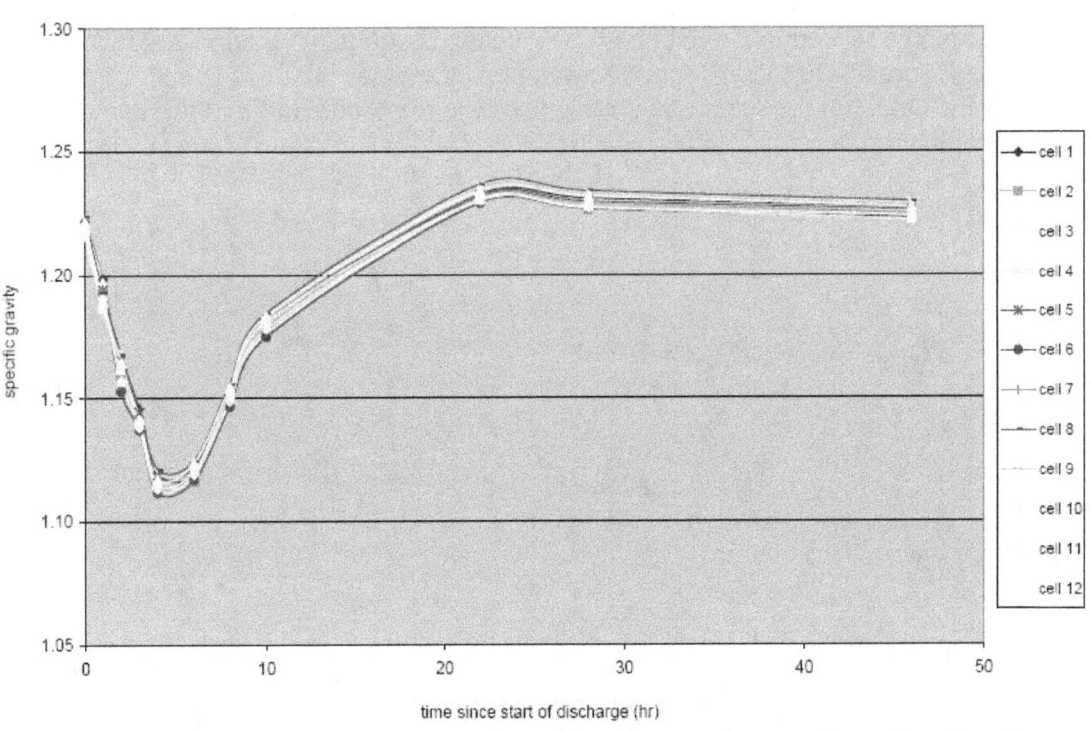

Figure 3-38   C&D cycle 10 midpoint specific gravity readings for all cells

3.4. Comparison of Float Current and Specific Gravity Responses

This section presents specific gravity and float current response data in a common graph to illustrate their response over time during the discharge and recharge cycles. The first example (Figure 3-39) shows the dramatic decrease in specific gravity (blue line) as measured at the midpoint of the cell followed by a steady increase to its original value during recharge.

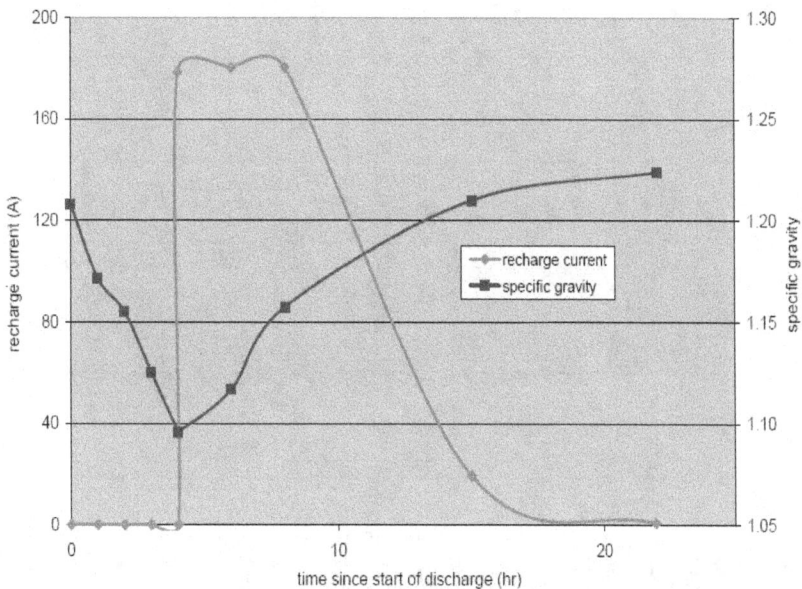

Figure 3-39   Enersys cycle 2 specific gravity vs. float current

The recharge current is depicted by the red line in Figure 3-39. With the battery charger disconnected during the performance test, the recharge current is zero. At approximately five hours following the initiation of the performance test, the recharge of the battery is initiated. The battery charger goes into a current limit mode (capped for this testing at 180 amps) and then decreases in an exponential manner towards a stable value of about 0.5 amps. Note the close relationship in time when both the specific gravity and the float current stabilize (approximately 20 hours).

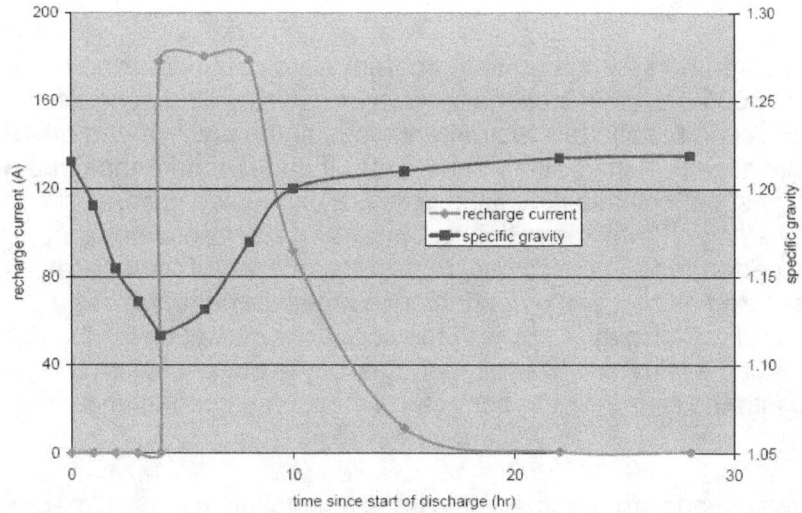

Figure 3-40   GNB cycle 4 specific gravity vs. float current

Figure 3-40 for the GNB battery is similar in overall appearance but reaches a stable value for both specific gravity and float current slightly sooner than the Enersys battery shown in Figure 3-39. Likewise, the largest of the three batteries, the C&D LCR-33 reaches stable specific gravity and float current values at approximately 25 hours as shown in Figure 3-41.

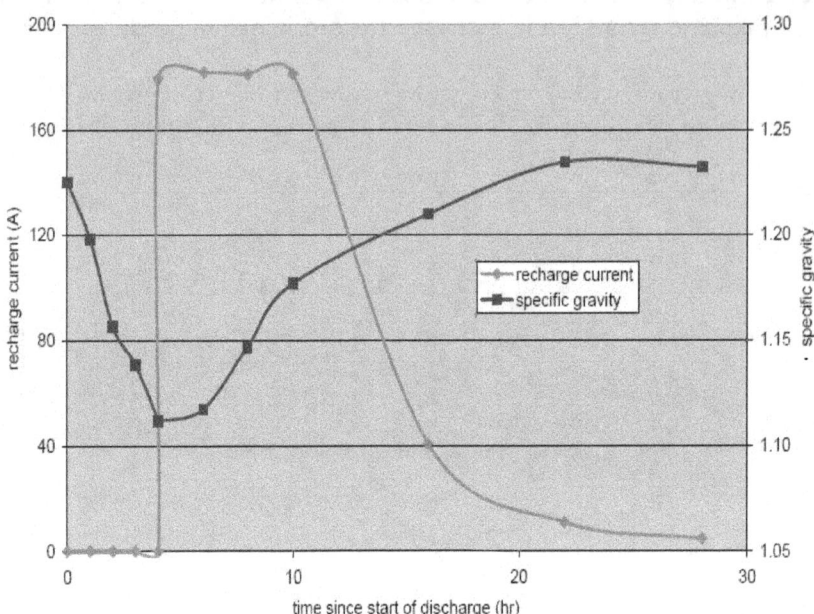

Figure 3-41   C&D cycle 8 specific gravity vs. float current

3.5. Conductance and Connection Resistance Data

In addition to the float current and specific gravity measurements discussed in the previous sections, individual battery cell conductance measurements were taken at key hold points during the performance of the battery discharge and recharge testing cycles. This section of the report describes the conductance and connection resistance data gathered during the battery performance testing cycles.

3.5.1.   Measurements

Conductance measurements for this testing program were obtained using the Midtronics Celltron Ultra CTU-6000 universal stationary battery analyzer. The conductance test is performed by connecting the two test instrument leads to the positive and negative terminal posts of the individual cells of the battery under test. If there is more than one positive and negative terminal post per cell and if multiple cells are connected in series by conductive straps to create a battery string or bank, as in this test program, a sequence of conductance measurements is performed. The electrical resistance of each of the cell-to-cell connecting straps is also measured as part of the conductance measurement process and is presented on the CTU-6000 LCD visual display along with the conductance values for the each of the battery cells in the string being evaluated. The temperature of the battery electrolyte must be measured and programmed into the instrument to correct the measured conductance values for thermal effects.

The measurement sequence for each specific battery installation can be preprogrammed into the instrument so that the operator can apply the probes to each of the cells in the battery-under-test, one-by-one, in the sequence required. The instrument LCD visual display will then

40

cue the operator to the next measurement in the sequence. When the measurement probes are applied to each of the testing points, a low-voltage, low-frequency ac signal is applied to the battery-under-test, and the instrument will measure the resulting ac current that flows in the test circuit. After the test circuit has stabilized (3-8 seconds was our experience during this test program), an audible tone sounds to alert the operator that the measurement step is complete and the next testing point in the sequence can begin; the results of the just-completed measurement step are presented on the LCD display along with a cue for the location of the next measurement point in the test sequence. A different audio tone is generated if a problem occurs during the performance of a conductance measurement step and information on the problem is shown on the instrument display. The technician can then make the necessary adjustments or corrections and then retry the measurement.

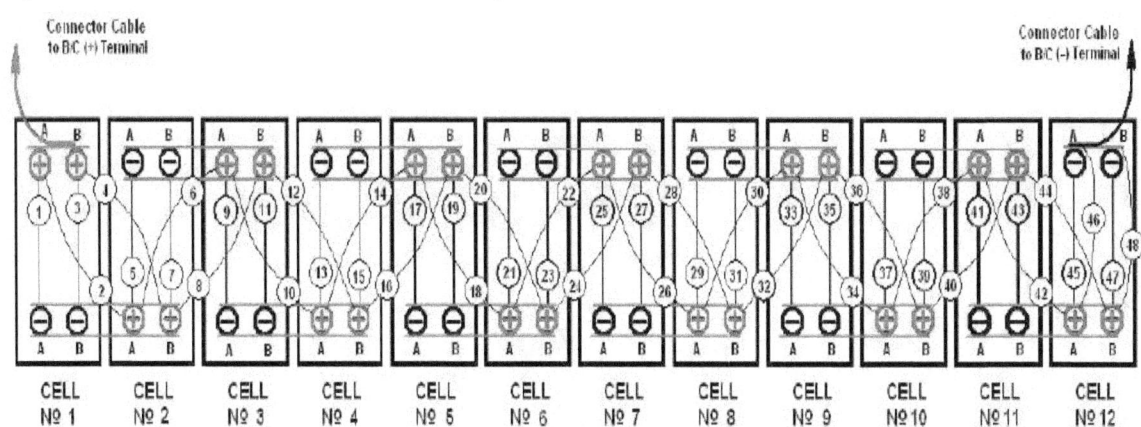

Figure 3-42   Conductance measuring sequence for the C&D LCR-33 battery string

Figure 3-42 shows the sequence of measurements required to measure the conductance for the C&D Model LCR-33, 12-cell, 4-posts-per-jar, single-tier rack-mounted Class 1E battery tested in this program. The results of the analysis provide the individual cell voltage and conductance, as well as the electrical resistance of the cell-to-cell connection straps, for the battery-under-test.

3.5.2.   Conductance and Connection Resistance Test Results

Conductance measurements were performed at three 'hold' points for each cell in a battery string as the battery was subjected to the 10 test cycles. The measurements were made as follows: 1) prior to the start of the initial four hour discharge, 2) after completion of the four-hour discharge but prior to start of the recharge, and 3) approximately 24 hours into the recharge. A fourth measurement was taken included at 168 hours after the start of the discharge test which, in fact, corresponded to the first measurement prior to the start of the initial four-hour discharge for the next test cycle.

Table 3-7 presents the averaged conductance measurements for the 12-cells in each battery string. As can be seen from the data, the conductance measured reflects the battery capacity at the time of the measurement: conductance measurements are highest in the fully charged state and lowest immediately at the end of the four-hour discharge segment of the performance test. The corresponding plots of the average conductance of the battery string show that the battery capacity is very nearly restored at the end of each step in the testing cycle.

41

Table 3-7   Conductance measurements – compared to battery capacity for all 30 cycles

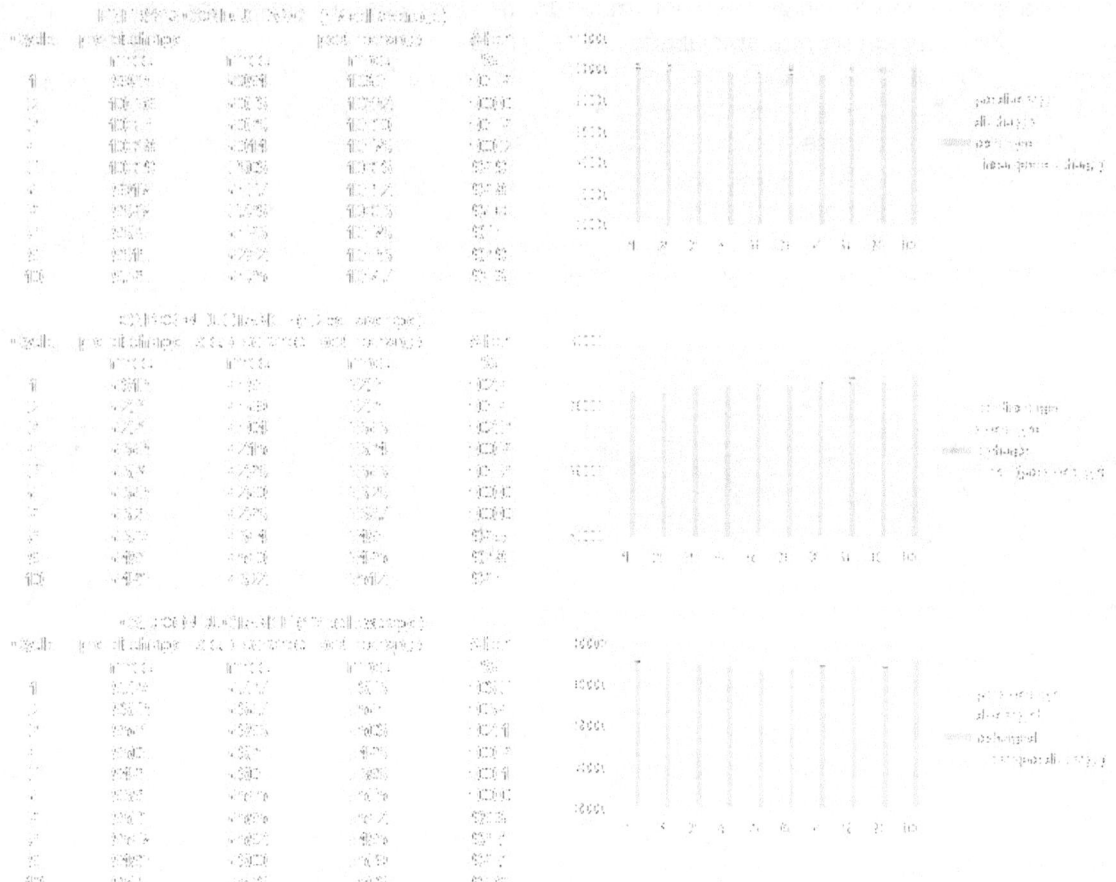

There is a trend of a slight decrease in the average fully-charged conductance for the Enersys GN-23 battery and the C&D LCR-33 battery over the course of the 10 testing cycles, but it is minimal. This corresponds to a similar indication of a trend showing slightly decreasing battery capacity as indicated by the last column in each table: battery capacity as measured by the Alber software. The GNB-Exide NCN-21 battery actually showed a trend to slightly increase its average fully-charged conductance measurement over the 10 test cycles. Battery capacity for the GNB-Exide NCN-21 as measured by the Alber software indicated a slightly decreasing battery capacity over the 10 cycles.

3.6. Temperature Monitoring Data

This section provides a summary of the temperature monitoring data accumulated during the discharge-recharge cycles. The significance of the temperature readings is that the performance test is temperature compensated based on the temperature at the start of the test. As can be seen in Figure 3-43, the average temperature of the cells from the start of the test to the completion of the test varies by approximately 8-10°F. We used the initial temperature only for setting the current to be used in the discharge test in accordance with IEEE Std. 450-2002. Slightly different capacity measurements would have resulted if we were to use an average temperature experienced over the course of the test.

While it is not practical to continuously compensate the data for temperature during the test, it is important to note that there is some impact to the battery capacity outcome based on the temperature compensation so the rationale employed for testing should be consistently applied.

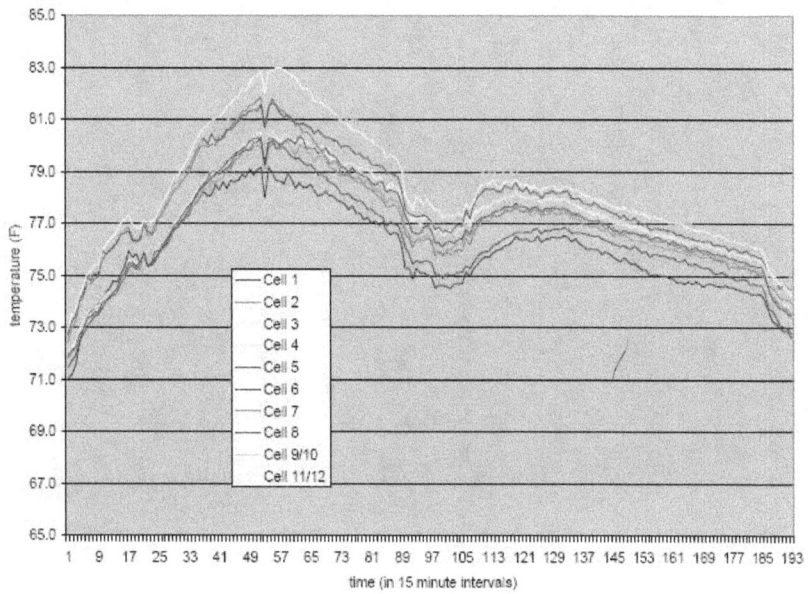

Figure 3-43   Enersys cycle 1 temperature response over time

This kind of response was also observed in the GNB and C&D battery strings as illustrated in Figure 3-44 (GNB cycle 5) and Figure 3-45 (C&D cycle 1). So despite the size of the battery jar or its ampere-hour rating, the surface temperature on the jar fluctuates in the same manner as the battery is discharged. It slowly returns to its initial condition as the battery is recharged. Ambient temperature was also monitored during the cycle testing. An example of the ambient change along with the comparable temperature changes for two of the cells is illustrated in Figure 3-46.

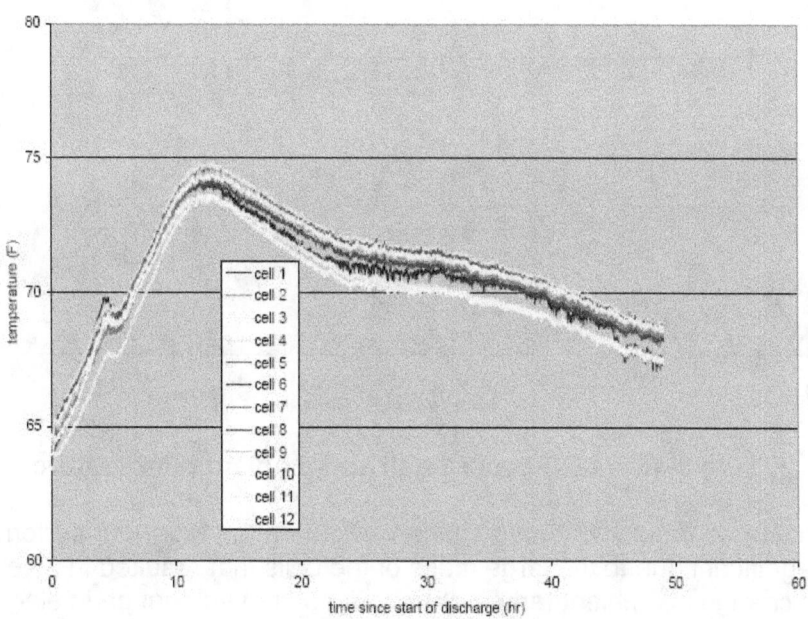

Figure 3-44   GNB cycle 5 temperature response over time

43

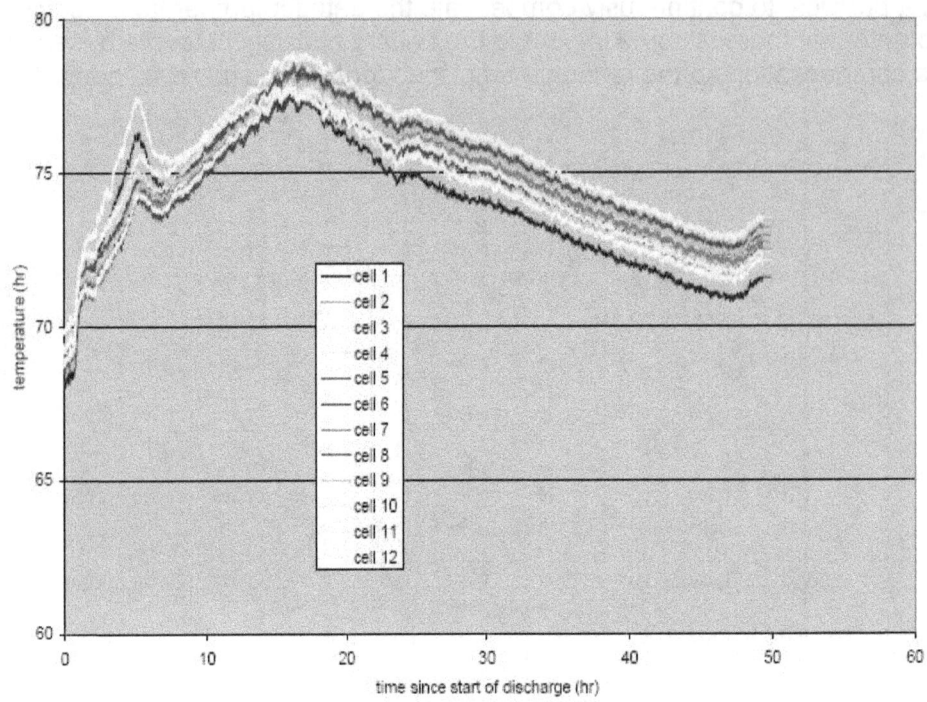

Figure 3-45   C&D cycle 1 temperature response over time

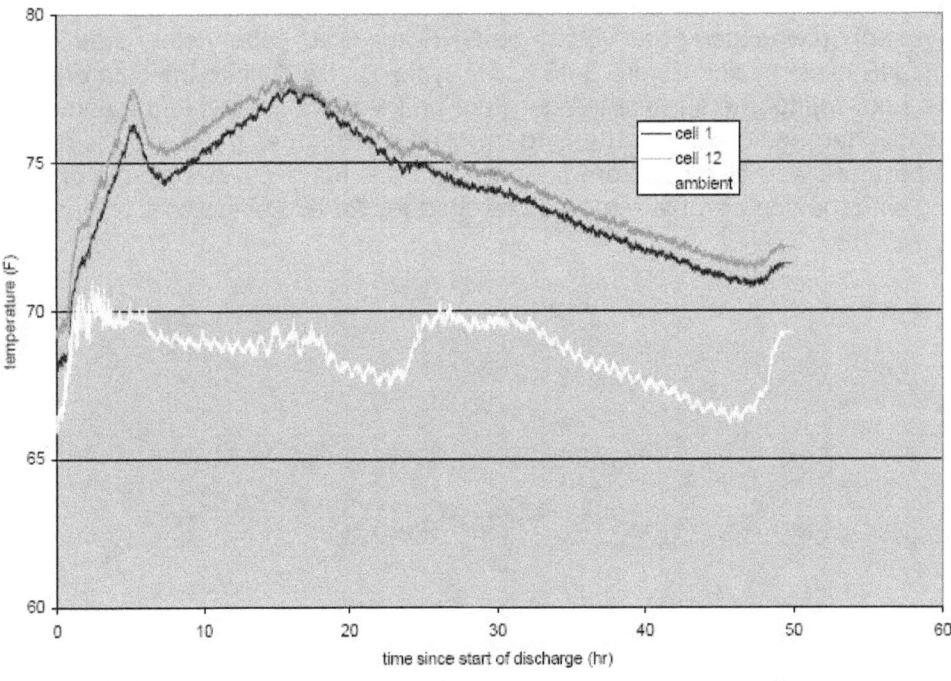

Figure 3-46   C&D cycle 1 cell and ambient temperatures

The slight ambient temperature fluctuation observed during the four hour performance testing was deemed insignificant due to the large mass of the cells that resulted in a several hour time delay between a change in ambient temperature to an observed change in electrolyte temperature. IEEE Std. 450-2002, Section 7.3.2.2 states that the temperature correction should be applied to a time based performance test in accordance with the formula, C= {Xa x K_C /Xt} x

100 where C is the % capacity at 25°C; Xa is the actual rate used for the test; Xt is the published rating for the time to reach a specified terminal voltage; and K_C is the correction factor for the cell temperature before the start of the test as provided in Table 2-2 of IEEE Std. 450-2002. For example, referring again to Figure 3-46, the rated four hour performance test for the C&D battery for cycle 1 was automatically corrected by the Alber Capacity test set so that the rating of 500 amps was corrected to 481 amps since our starting electrolyte temperature was 70°F (K_C = 1.04).   In the actual case of cycle 1 for the C&D battery, it took four hours and 15 minutes to reach an overall string voltage of 21.0 volts. This yielded a capacity of 106.5% for cycle 1.

Figure 3-47 illustrates the location of the surface mounted thermocouple for one of the batteries.

Figure 3-47   Surface Mounted Thermocouple

3.7.   Summary

In this section, we have summarized the extensive amount of data that were acquired during the 30 cycles of discharge and recharge testing. Specific gravity, charging (float) current, conductance, and temperature were frequently measured during the cycles. Calibration checks were performed on the equipment periodically to ensure the quality of the data. Regular inspections of the batteries, including electrolyte level and connection resistance, ensured that the batteries were tested within their normal operating limits.

Both recharge/float current and specific gravity responded similarly during the charging cycle indicating that either can provide a measure of a battery's state-of-charge. The three battery types responded very similarly during the discharge and recharge cycles lending a level of confidence to generalizations made that can be used in regulatory guidance. The analyses and conclusions are discussed further in Sections 4 and 5, respectively.

4.   ANALYSIS OF TEST RESULTS

In this section of the report, further analyses of the data are discussed in the following areas:

- Float current as a measure of "state-of-charge"
- Specific gravity as a measure of "state-of-charge"
- Capacity testing trends and their meaning,
- Electrolyte stratification effects, and
- "Return-to-service" test results.

4.1.   An Assessment of Float Current as a Measure of State-of-Charge

Specific gravity has been used for many years as a means of determining a battery's state-of-charge. For the lead-calcium type of vented lead acid battery used in nuclear power plants, there are often technical specification requirements associated with specific gravity, such as frequency of checking, the expected values to be obtained, and the corrective action, or limiting conditions for operation, that are imposed if specific gravity readings are lower than required. In addition, battery manufacturers include specific gravity measurements as a standard maintenance check for their batteries.

One of the changes made over the years and supported in IEEE Standard 450-2002, is the use of charging/float current as a means to monitor the state-of-charge of lead acid batteries. Revision 2 of Regulatory Guide 1.129, issued in February 2007, endorsed IEEE Std. 450-2002 with certain clarifying regulatory positions that are described in Section C of the Regulatory Guide.

In this testing program, we calculated the ampere-hours that are restored to the battery as it was recharged as one means to determine that the battery had been recharged. In many test cycles, the total number of ampere-hours returned to the battery exceeded the number of ampere-hours discharged <u>before</u> the float current reached a steady-state level. For the lead acid batteries that we tested, the stable float current achieved was in the 0.5 to 2.0 amp range. At that point in the current vs. time curve, only small numbers of ampere-hours are being returned to the cell. This is illustrated in Figures 4-1 to 4-5 for the three different battery strings. In Figure 4-1, the float current response below 10 amps is shown for the cycle 10 test of the Enersys battery. Figures 4-2 and 4-3 depict the float current response for the GNB battery when the recharge was conducted at 27.0 and 28.0 volts, respectively. Similarly, Figures 4-4 and 4-5 illustrate the response for the C&D battery with the recharge conducted at 27.0 and 28.0 volts, respectively.

Note that the largest of the cells, the C&D LCR-33 model, took a longer time to reach a stable float current due to the higher number of ampere-hours that needed to be returned to the battery following the performance test. The battery charger that was used in the project was limited to 180 amps (current limit setting). As shown in Figure 4-5, the charger output switched from 28 volts to 27 volts at 40 hours, thereby resulting in a corresponding decrease in the float current at that point.

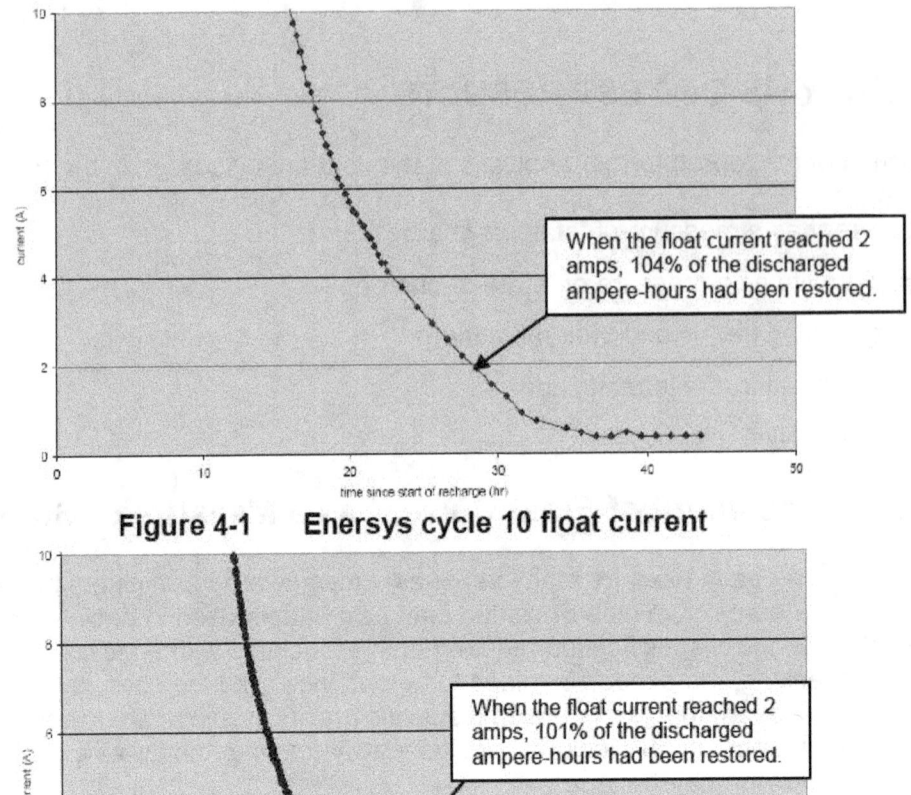

Figure 4-1   Enersys cycle 10 float current

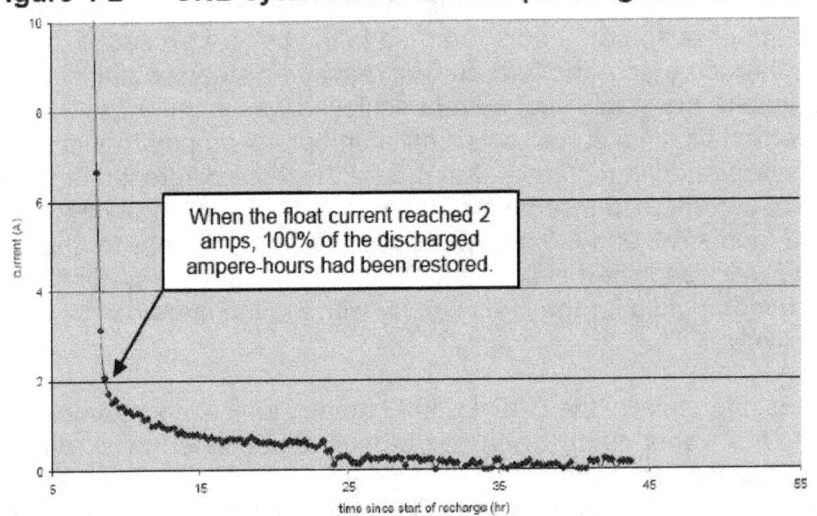

Figure 4-2   GNB cycle 9 float current (recharge at 27.0 volts)

Figure 4-3   GNB cycle 1 float current (recharge at 28.0 volts)

48

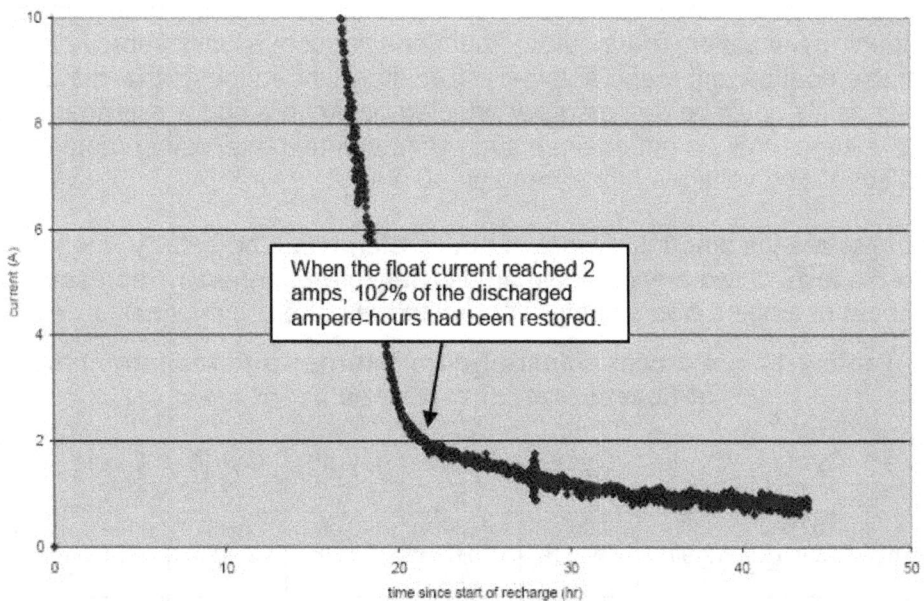

Figure 4-4   C&D cycle 2 float current (recharge at 27.0 volts)

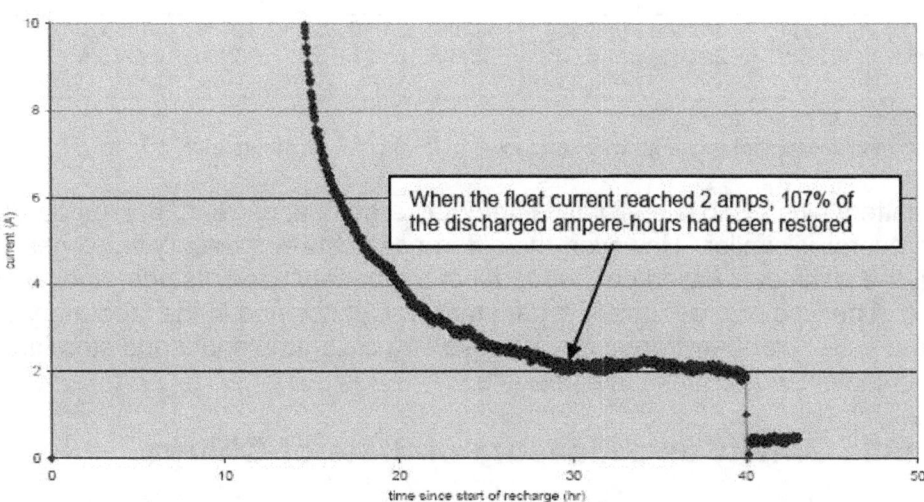

Figure 4-5   C&D cycle 10 float current (recharge at 28.0 volts for 40 hours)

The float current responses on the three battery strings support the following conclusions:

1. The battery's state-of-charge as measured by the return of the discharged ampere-hours to the battery can be correlated to the stabilizing of the float current (generally about 0.5 to 2.0 amps). The stable current is slightly higher when the applied voltage (equalizing charge) is higher.

2. Once the float current reaches an asymptotic level, the ampere-hours being returned to the battery are minimal. This observation relates to the statement in section C.3 (d) of Regulatory Guide 1.129, Revision 2 that recommends a stable float current reading be achieved for three hours before the battery is considered to be near full charge. Testing to confirm that this approach is valid is discussed further in Section 4.6, "Return-to-Service" Tests.

3. Recharging at an equalizing voltage rather than a float level voltage results in returning ampere hours to the battery more quickly, but does not consistently improve the time to reach a stable float current (see Table 4-1). The literature suggests that the condition of the battery over its life could be detrimentally affected by recharging for extended times at an equalizing voltage. We did not observe such effects during this testing program under controlled equalizing voltages (not exceeding 40 hours).

Table 4-1 summarizes the amount of ampere-hours returned to the battery when the float current reached a level of two amps. A two amps float current generally represents the point where the amount of ampere-hours being returned to the battery is minimal.

**Table 4-1   Percent ampere-hours returned and recharge time
at float current of two amps**

Battery Type	Cycle 1	Cycle 2	Cycle 3	Cycle 4	Cycle 5	Cycle 6	Cycle 7	Cycle 8	Cycle 9	Cycle 10
Enersys	101% 13.0 h	102% 15.9 h	102% 18.9 h	103% 19.9 h	103% 20.6 h	104% 22.2 h	104% 26.0 h	104% 26.3 h	103% 26.8 h	104% 28.6 h
GNB	100% 8.5 h	102% 8.8 h	103% 14.8 h	101% 15.8 h	101% 16.6 h	*	101% 19.0 h	102% 19.7 h	101% 19.7 h	101% 19.8 h
C&D	102% 20.6 h	102% 21.3 h	105% 24.0 h	105% 24.0 h	105% 25.5 h	103% 28.6 h	103% 30.2 h	103% 31.3 h	107% 40.0 h	107% 38.7 h

*data unavailable due to power interruption
Note: shaded cycles indicate that recharge was conducted at 28.0 volts; all others at 27.0 volts

While each battery type exhibits a slightly different charge/float current, their fundamental performance overall is similar. However, the differences between battery types need to be reconciled at the plant level based on the model of the battery used, its age, and its manufacturer. The type of float current testing that was performed in the laboratory can be conducted in a power plant environment. The use of a calibrated shunt and standard data acquisition equipment would make this possible.

4.2. Specific Gravity as a Measure of State-of-Charge

As discussed in Section 3, midpoint specific gravity responds very similarly to charging/float current during battery recharge. That is, it reaches a stable value in the same manner as the charging/float current. Figures 4-6 to 4-8 illustrate this relationship for each battery type.

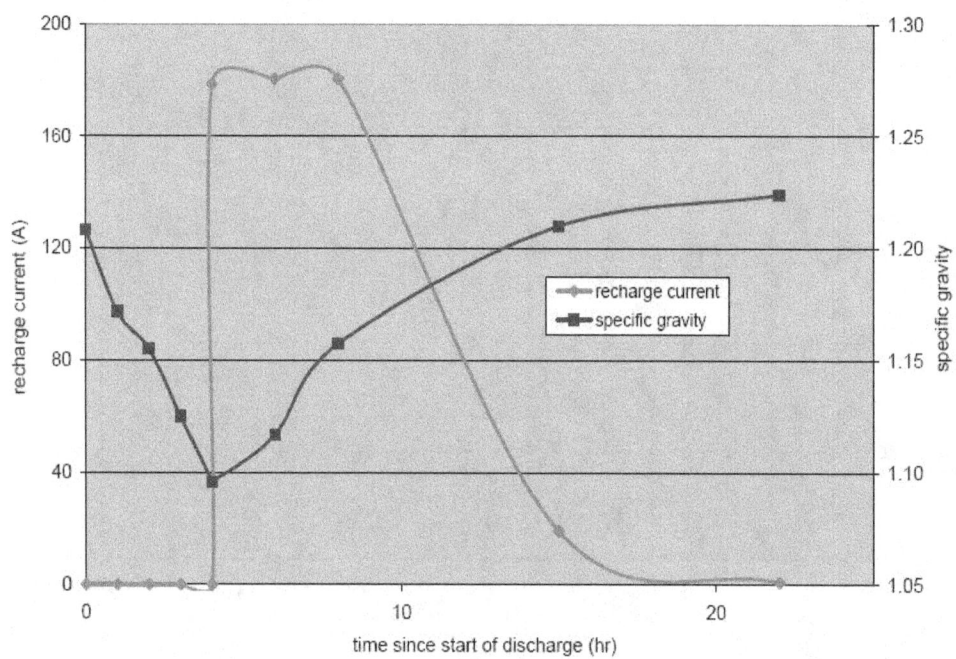

Figure 4-6   Enersys cycle 2 specific gravity vs. float current

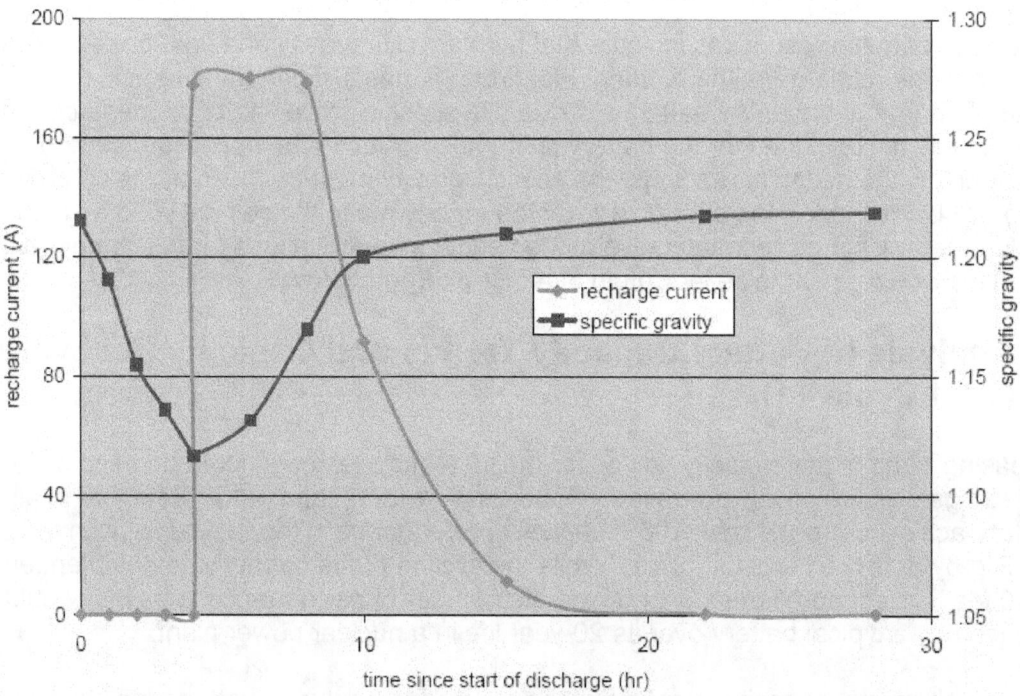

Figure 4-7   GNB cycle 4 specific gravity vs. float current

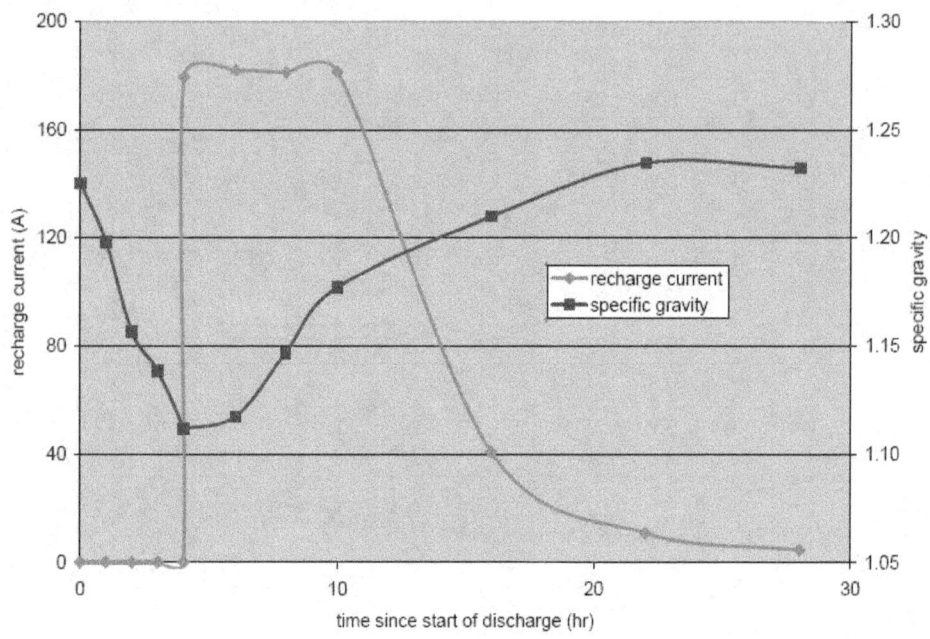

Figure 4-8   C&D cycle 8 specific gravity vs. float current

The analyses of these responses indicate that both specific gravity and float current are equally responsive to the recharge of the battery. Float current has a slight advantage in that it provides the state of charge of the entire battery string, while specific gravity must be measured for a number of cells to conclude that the battery string has achieved the necessary state-of-charge. When using just the midpoint readings, the specific gravity must be taken at the correct depth within the cell. Measuring the specific gravity too far down into the cell will yield an artificially high value that will not be representative of the vendor's specifications for full charge (typically 1.215). This would give a false indication of a fully charged battery.

4.3.   Analysis of Battery Capacity Trends and Capacity Recovery (Test Cycle 11)

A decreasing trend of the capacity values for the 10 test cycles conducted on each of the battery strings was previously portrayed in Figures 3-11, 3-12, and 3-13. Note that in all cases, the battery achieved greater than 100% capacity performance in the first cycle, with the capacity decreasing by 5-10% by the 10th cycle. While the cycling of the battery was not intended to age it, the 10 test cycles represents the approximate number of performance tests that would be experienced by a typical battery over its 20-year life in a nuclear power plant.

Initially, the change in capacity over time was assumed to be due to battery depletion as it was exercised during the deep cycle testing. However, it was also noticed that the amount of electrolyte stratification increased over the 10 cycles. The question was raised whether this increase in stratification could be the cause of the capacity decrease, mainly due to the low specific gravity measured at the top of the cell. The following paragraphs provide analysis of the change in specific gravity and the corresponding change in capacity that occurred over the first 10 cycles of testing, and describes the results of the performance test that was performed (cycle 11) for each battery string several months after the electrolyte had reached an equilibrium condition.

For the Enersys battery, the capacity changed from 101.7% to 94.3%, a decrease of 7.4% over the 10 cycles. As shown in Figure 4-9, the specific gravity prior to the start of the cycle for the midpoint reading changed from 1.215 in cycle 1 to 1.185 at the beginning of cycle 10, a 2.5% change, while the top reading changed from 1.217 prior to the beginning of cycle 1 to 1.113 prior to the beginning of cycle 10, an 8.5% decrease. The bottom reading changed from 1.218 to 1.31, an increase of 9.2%.

To help determine the impact of the electrolyte stratification on battery capacity, the NRC project manager approved the running of an eleventh four hour performance test after this battery had been on a float charge for 10 months. The results were that the capacity recorded was 104.7%, even higher than the when the battery string was received. The capacity for the Enersys battery over the 11 test cycles is illustrated in Figure 4-10.

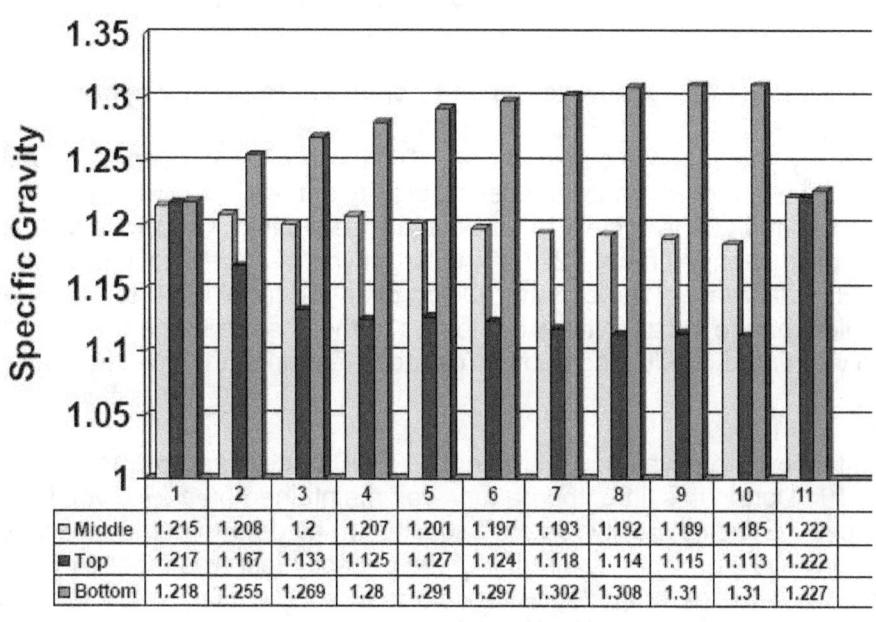

	1	2	3	4	5	6	7	8	9	10	11
□ Middle	1.215	1.208	1.2	1.207	1.201	1.197	1.193	1.192	1.189	1.185	1.222
■ Top	1.217	1.167	1.133	1.125	1.127	1.124	1.118	1.114	1.115	1.113	1.222
▤ Bottom	1.218	1.255	1.269	1.28	1.291	1.297	1.302	1.308	1.31	1.31	1.227

Test Cycle

Figure 4-9   Specific gravity prior to the start of each test cycle - Enersys battery

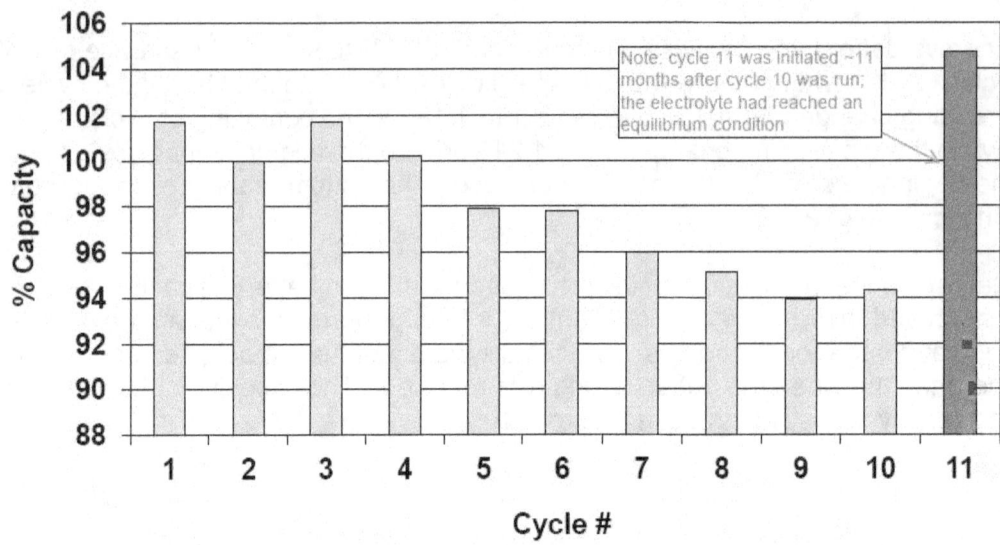

Figure 4-10   Enersys capacity changes including cycle 11

For the GNB battery, the capacity changed from 102.4% to 97.1%, a decrease of 5.3% over the 10 cycles. As shown in Figure 4-11, the specific gravity at the start of the cycle for the midpoint reading changed from 1.209 in cycle 1 to 1.206 at the beginning of cycle 10, virtually no change, while the top reading changed from 1.217 at the beginning of cycle 1 to 1.126 at the beginning of cycle 10, a 7.5% decrease. The bottom reading changed from 1.212 to 1.283, an increase of 5.9%. The smaller change in capacity as compared to the change in the top specific gravity readings again warranted conducting another capacity test once the specific gravity had reached equilibrium.

Cycle 11 of the test program was conducted on the GNB battery approximately eight months following cycle 10. During this time, the battery was maintained on a 27.0 volt float voltage (2.25 volts per cell). Specific gravity reached equilibrium as determined by the weekly profile readings taken on cells 9 and 11. The result of the capacity test was that 101.8% capacity was obtained, nearly the same as when the battery string was fresh. The 11 cycles of testing are depicted in Figure 4-12.

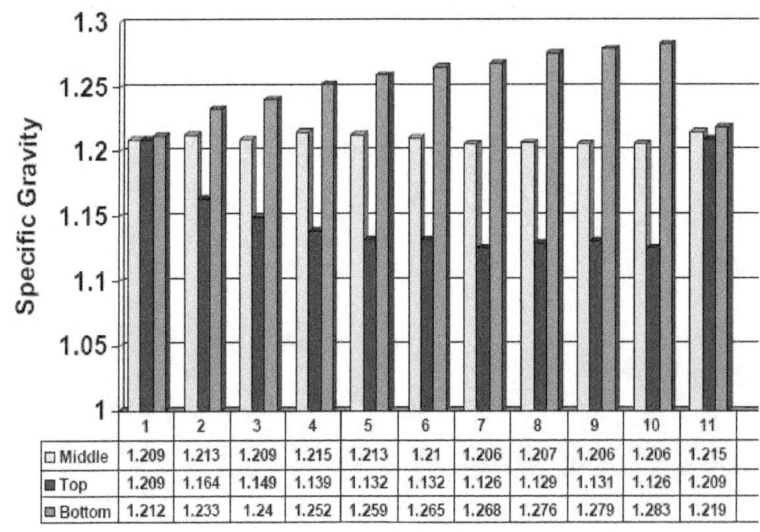

	1	2	3	4	5	6	7	8	9	10	11
□ Middle	1.209	1.213	1.209	1.215	1.213	1.21	1.206	1.207	1.206	1.206	1.215
■ Top	1.209	1.164	1.149	1.139	1.132	1.132	1.126	1.129	1.131	1.126	1.209
▨ Bottom	1.212	1.233	1.24	1.252	1.259	1.265	1.268	1.276	1.279	1.283	1.219

Test Cycle

Figure 4-11   Specific gravity at the start of each test cycle - GNB

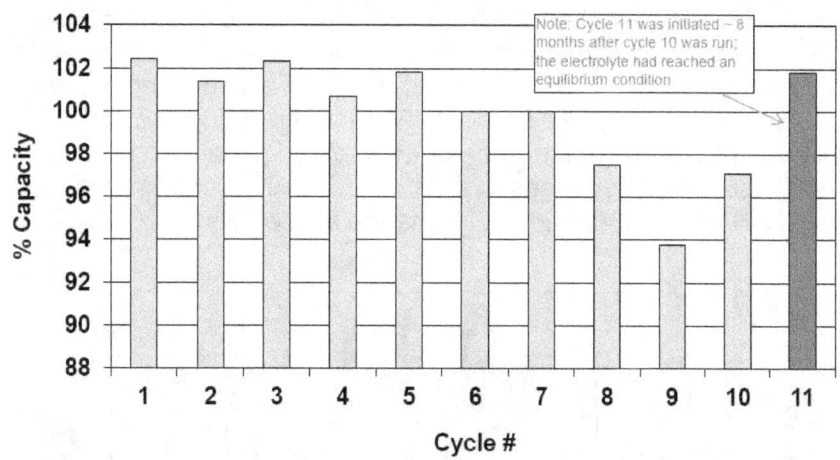

Figure 4-12   GNB capacity changes including cycle 11

For the C&D battery, the capacity changed from 106.5% in cycle 1 to 96.6% in cycle 10, a decrease of 9.9%. As shown in Figure 4-13, the specific gravity at the start of the cycle for the midpoint reading changed from 1.218 in cycle 1 to 1.22 at the beginning of cycle 10, virtually no change, while the top reading changed from 1.209 at the beginning of cycle 1 to 1.126 at the beginning of cycle 10, a 6.8% decrease. The bottom reading changed from 1.212 to 1.283, an increase of 5.8%. These smaller changes in the profile readings and, therefore, lesser stratification, should not result in a nearly 10% change in capacity.

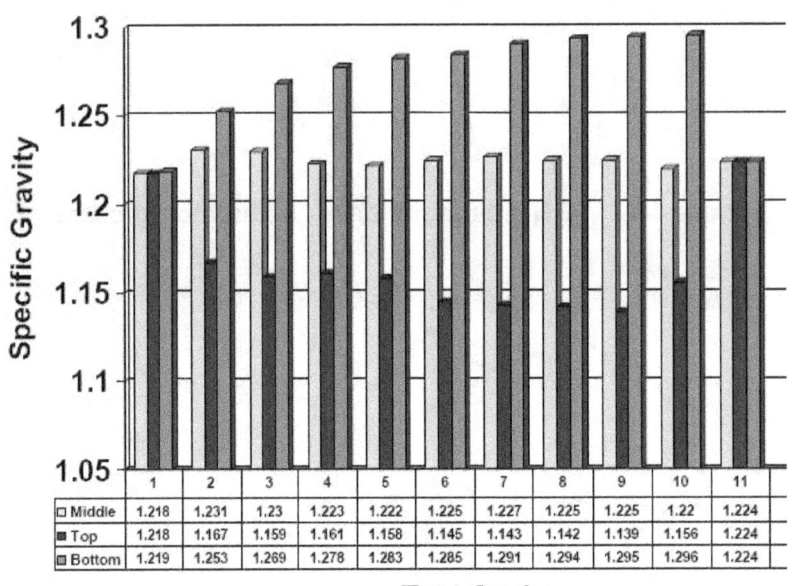

	1	2	3	4	5	6	7	8	9	10	11
☐ Middle	1.218	1.231	1.23	1.223	1.222	1.225	1.227	1.225	1.225	1.22	1.224
■ Top	1.218	1.167	1.159	1.161	1.158	1.145	1.143	1.142	1.139	1.156	1.224
▣ Bottom	1.219	1.253	1.269	1.278	1.283	1.285	1.291	1.294	1.295	1.296	1.224

Test Cycle

Figure 4-13   Specific gravity at the start of each cycle - C&D

Cycle 11 testing was performed approximately 5.5 months after the completion of cycle 10. The electrolyte stratification observed at the time of cycle 10 had slowly over time disappeared while the battery was maintained on a 27.0 volt float voltage (2.25 volts per cell). With the electrolyte at equilibrium, the capacity returned to 102.3% as indicated by Figure 4-14.

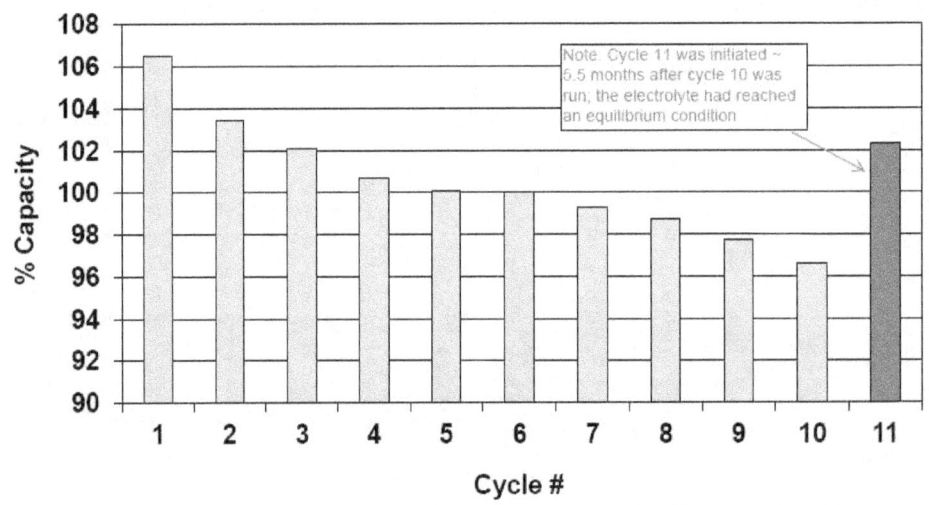

Cycle #

Figure 4-14   C&D capacity changes including cycle 11

The above data suggest that electrolyte stratification has an impact on battery capacity. This is probably true to some extent, but other factors must be considered as well. The first is that the scheduling of weekly capacity tests on the battery string did not allow the full recovery of battery capacity, a phenomenon sometimes called "capacity walkdown." The second is that by maintaining the battery on float for several months following the deep-cycle testing, we were slowly but steadily putting ampere-hours back into the cells while at the same time allowing the electrolyte to equilibrate.

We note that one of the battery vendors (C&D) states in their operating manual (RS-1476, Section 4) that typically a standby battery like the one we tested "will not be subjected to more than one test discharge each year" and "Additional deep and/or frequent discharges can shorten service life, even with proper maintenance and operation". It is therefore unlikely that nuclear power plants will experience as significant a degree of electrolyte stratification as we did as a result of the ten consecutive weeks of discharge tests. However, even after a single discharge test, nuclear power plants will experience some electrolyte stratification, and must therefore be cognizant of how specific gravity measurements are taken to ensure a representative value is obtained.

4.4. Use of Conductance for Monitoring Battery Condition

There is a trend of a slight decrease in the average fully-charged conductance for the Enersys GN-23 battery and the C&D LCR-33 battery over the course of the 10 test cycles, but it is only minimal. This corresponds to a similar indication of a trend showing slightly decreasing battery capacity as indicated by the last column in each table: battery capacity as measured by the Alber software. The GNB-Exide NCN-21 battery actually showed a trend to slightly increase its average fully-charged conductance measurement over the 10 test cycles. Battery capacity for the GNB-Exide NCN-21 as measured by the Alber software indicated a slightly decreasing battery capacity over the 10 cycles.

Figures 4-15 to 4-17 illustrate the conductance trends for sample cells in each of the three battery types. The Enersys cell conductance appears to follow a similar trend to the measured battery capacity. Conductance would be a useful measure if in fact this correlation was consistent and true for all battery types. The samples chosen from the GNB and C&D battery strings do not show a pronounced change over the 10 cycles of performance testing. While the absolute values of conductance vary for the GNB and C&D batteries, they both show quite a large variation among the cells (~1000 mhos); however, the cells operated uniformly by all other measures. We, therefore, cannot draw any conclusions about the overall usefulness of conductance from these series of tests.

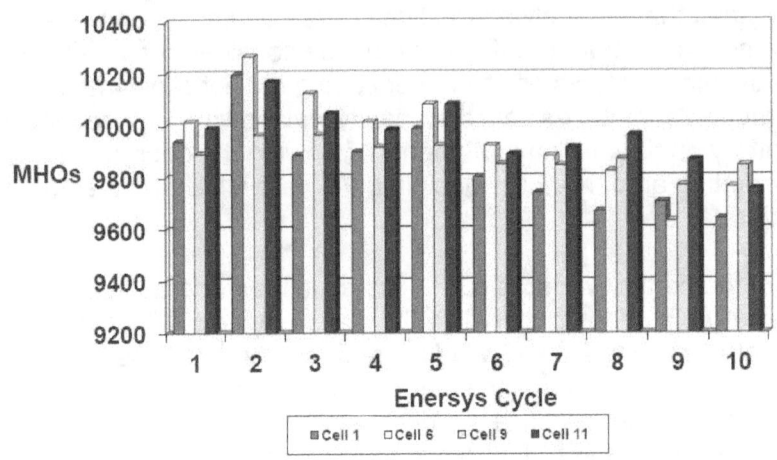

Figure 4-15   Sample cell conductance pre-test measurements - Enersys

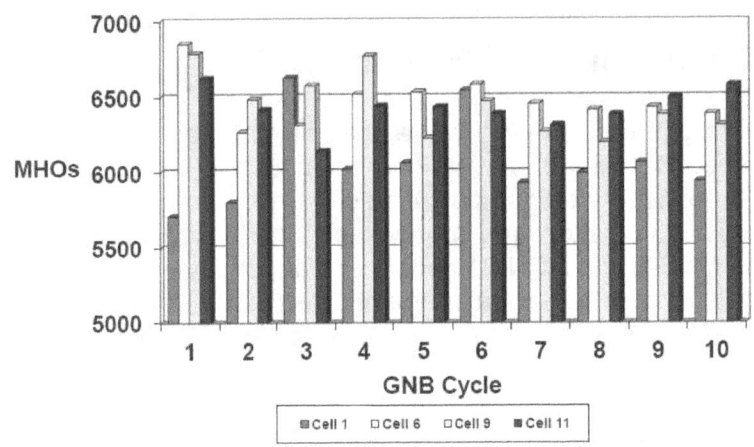

Figure 4-16   Sample cell conductance pre-test measurements - GNB

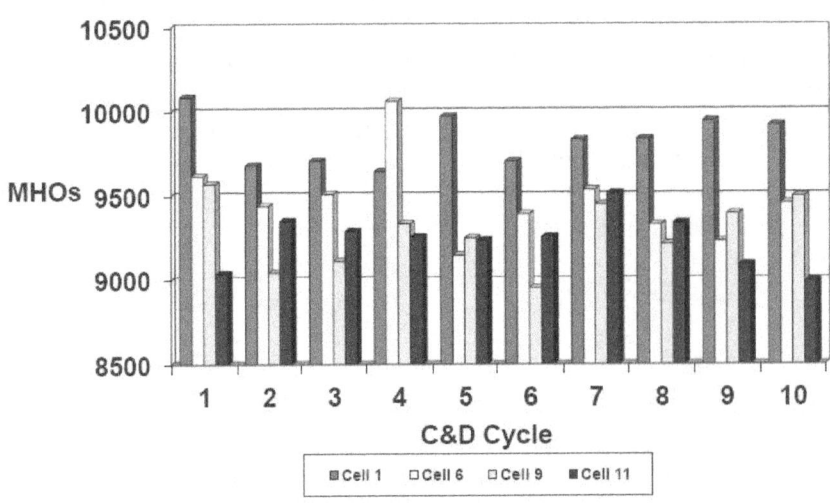

Figure 4-17   Sample cell conductance pre-test measurements - C&D

58

While these data might be useful in determining if a particular cell is experiencing degradation, we found it less useful for ascertaining the overall state-of-charge of the battery string. As stated in Section 3, the conductance meter we used also had the ability to measure connection resistance which we found useful as part of our battery operability checks. Annual checks of connection resistance (cell-to-cell and terminal) are recommended by IEEE Std. 450-2002, Section 5.2.3.

4.5. Analysis of Temperature Measurements

IEEE Std. 450-2002 devotes a fair amount of discussion to the impact of temperature on battery performance, including the following:

- Monthly checks of ambient temperature; quarterly checks of the temperature of 10% of the battery cells.

- For selecting the discharge rate for a capacity test, the Standard states that for the time adjusted method, the temperature correction should be applied to the capacity calculations. Table 1 in the Standard provides the recommended temperature correction factor.

- In Annex B.2, the Standard states that the specific gravity values are based on a temperature of 77°F, and should be corrected for the actual electrolyte temperature.

- Annex C.4 states that the internal resistance decreases and the electrochemical reaction rates increase as the temperature of the electrolyte increases. This section of the Standard refers to Annex D.3 which states that large cell temperature deviations are sometimes caused by shorting conditions, which are also evident by abnormal cell voltage and/or increasing float current. This same section indicates that operation of the battery at elevated temperatures will reduce life expectancy but will not adversely affect battery capacity. Annex H describes this effect in more detail.

Figure 4-18 is provided as an example of the temperature response observed during the testing as measured by surface-mounted thermocouples. In this example the temperature change in some cells ranges from about 68°F to 77°F during the course of the 4-hour performance test. The load to be applied for the performance test is temperature compensated based only on the initial temperature of the electrolyte. It is not practical to change the temperature compensation dynamically during the conduct of the test, however, the cell temperature response is worthy of note if only to demonstrate that during high load conditions like a 4-hour performance test, a substantive increase in electrolyte temperature should be expected.

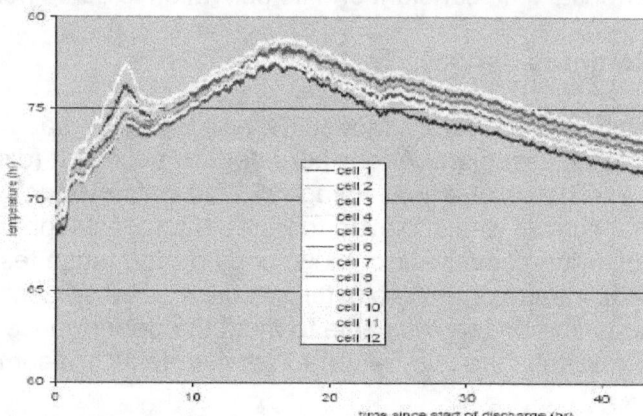

Figure 4-18   Analysis of temperature impact on capacity tests (C&D Cycle 1)

4.6. "Return-to-service" Tests (Test Cycles 12 to 14)

The purpose of this testing was to evaluate the ability of a battery to meet its capacity and capability requirements when it has met the "return-to-service" criteria. The "return-to-service" terminology is common in nuclear power plant technical specifications when referring to equipment operability, but it is also inferred in IEEE Std. 450-2002 (Annex A.2) in the discussion associated with determining the fully charged condition of a battery. In this latter context, IEEE Std. 450-2002 states that "When the charging current has stabilized at the charging voltage for three consecutive hourly measurements, the battery is near full charge." The nuclear industry and the U.S. NRC, through Technical Specification Task Force (TSTF) - 500, have been working to revise the technical specifications to reflect the use of float current as a means of determining battery operability. It is expected that the results of this testing can be used to help demonstrate the point at which a battery can be returned to service and be expected to fulfill its safety function.

In this testing program, the first criterion used to determine the "return-to-service" time was when a stabilized recharge float current was achieved for three-hours. When the battery had been recharged to that point, a second performance test was conducted to determine that the battery could meet its design function (>80% capacity). Following this second performance test, the battery string was recharged for at least 72 hours before the test was repeated.

The second criterion used to determine the "return-to-service" time was when the float current reached a value equivalent to three time constants on the exponential shaped recharge/float current curve. For our battery strings, this occurred when the float current reached 9 amps, as compared to between 0.5 and 2 amps for achieving a stable float current. At that point, the recharge was terminated and a second performance test was conducted. In both cases, the battery strings were successful in meeting their design function as described in this section.

The utility of the recharge time constant concept in evaluating the state-of-charge and the "return-to-service" limits for a battery is evident from the testing performed. Each battery type has its own unique characteristics and the limits for evaluating the state-of-charge and the "return-to-service" limits for a battery would have to be established on a case-by-case basis that includes knowing the battery charger recharge parameters such as its current limit setting and its recharge voltage value. However, as discussed later in this section, the recharge time constant concept provides an opportunity to reconsider the protocols whereby decisions are made concerning the operability status of batteries when returning them to service. A detailed explanation of the exponential time constant and its derivation is contained in Appendix A.

4.6.1. Enersys "Return-to-service" Tests

"Return-to-service" Testing of the Enersys Model 2GN-23 Batteries was conducted in accordance with an approved test plan. A four-hour performance test (discharge test) was conducted followed by a recharge at a float voltage of 27.0 volts. When the battery reached the point where a stable float current was achieved for three hours, a second four hour performance test was conducted. Following completion of the second performance test, the battery was again recharged at a float voltage of 27.0 volts for five days. The process was then repeated a second time. In both tests, the battery was able to meet the performance requirements (>80% of rated capacity). The results of the two "return-to-service" tests are summarized below.

Enersys Cycle 12: The initial performance test for cycle 12 resulted in a capacity of 101.0%. The recharge of the battery was initiated and reached a float current of 1.1 amps where it remained stable (within 0.5 amps) for the next three hours. At this time the charger output was secured and a series of specific gravity measurements were taken. The second performance test was initiated approximately 24 hours from the start of the recharge. The results from this second capacity test revealed a battery capacity of 98.6%. Following a series of specific gravity measurements, the battery was placed on a 27.0 volt float charge. Figure 4-19 illustrates the float current response at the point of current stabilization for cycle 12.

Enersys Cycle 13: The initial performance test for cycle 13 resulted in a capacity of 96.9%. The recharge of the battery was initiated and reached a float current 1.0 amp where it remained stable (within 0.5 amps) for the next three hours. At this time the charger output breaker was opened and a series of specific gravity measurements were taken. The second performance test was initiated approximately 26 hours from the start of the recharge. The results from this second capacity test revealed a battery capacity of 94.8%. Following a series of specific gravity measurements, the battery was placed on a 27.0 volt float charge. Figure 4-20 illustrates the float current response at the point of current stabilization for cycle 13.

Analysis of Enersys "Return-to-service" Testing – Stable float current: When the Enersys battery was returned to service following a four-hour performance test, it was successful in two tests (Enersys cycles 12 and 13) in demonstrating that it could meet its performance requirements when recharged to the point where the float current was stable for three hours. For stable current, we used the criteria that the float current remained the same (within 0.5 amps) for three consecutive hours. In both cases this occurred when the float current was about 1.0 amp.

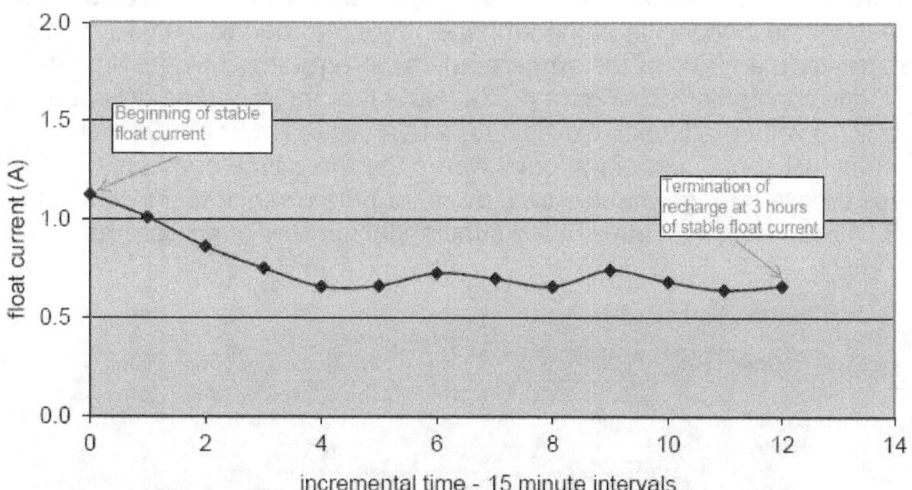

Figure 4-19   Enersys cycle 12 stable float current monitoring

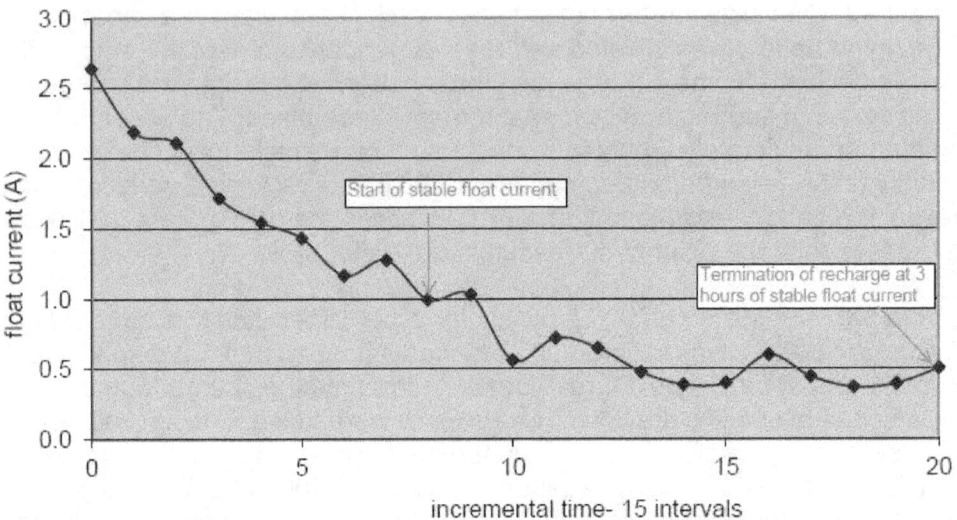

Figure 4-20   Enersys cycle 13 stable float current monitoring

Enersys Cycle 14: An initial performance test for cycle 14 was conducted as in cycles 12 and 13. The recharge of the battery was initiated following the completion of a set of specific gravity readings. The battery reached a float current of 9 amps (three time constants) at approximately 12.5 hours from the start of the recharge (see Figure 4-21). At this time the charger output was secured and a series of specific gravity measurements were taken. The second performance test achieved a battery capacity of 93.3%. Following a series of specific gravity measurements, the battery was placed on a 27.0 volt float charge.

Analysis of Enersys "Return-to-service" Testing - Three Time Constants: During the first 10 cycles of testing, the calculation of the ampere-hours returned to the battery during recharge indicated that greater than 90% of the ampere hours were returned by the time the float current reached three time constants (see Figure 4-22). Note that the recharge was conducted at a float voltage of 27.0 volts (2.25 volts per cell) for all 10 cycles. In Enersys Cycle 14, the four hour performance test conducted at the point where the three time constants had been obtained confirmed these results through the use of a four hour performance test in which the battery achieved a 93.3% capacity, indicative of the substantial number of ampere-hours returned to the battery at that point.

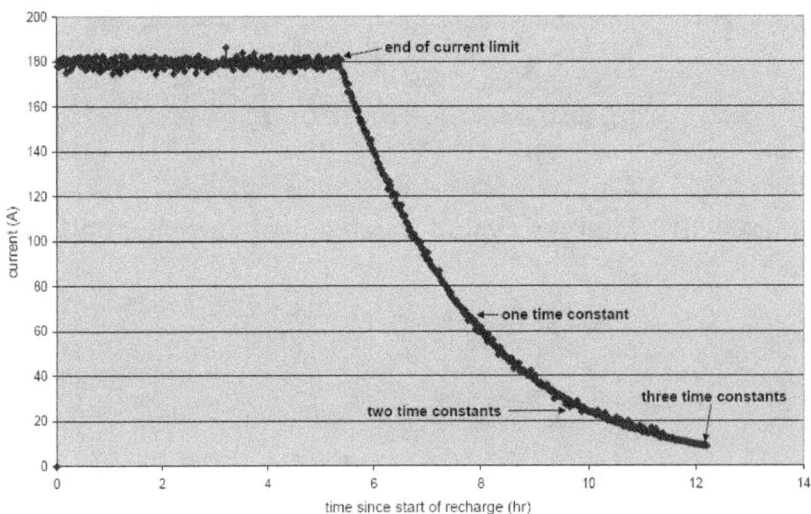

Figure 4-21   Enersys cycle 14 - float current to three time constants

62

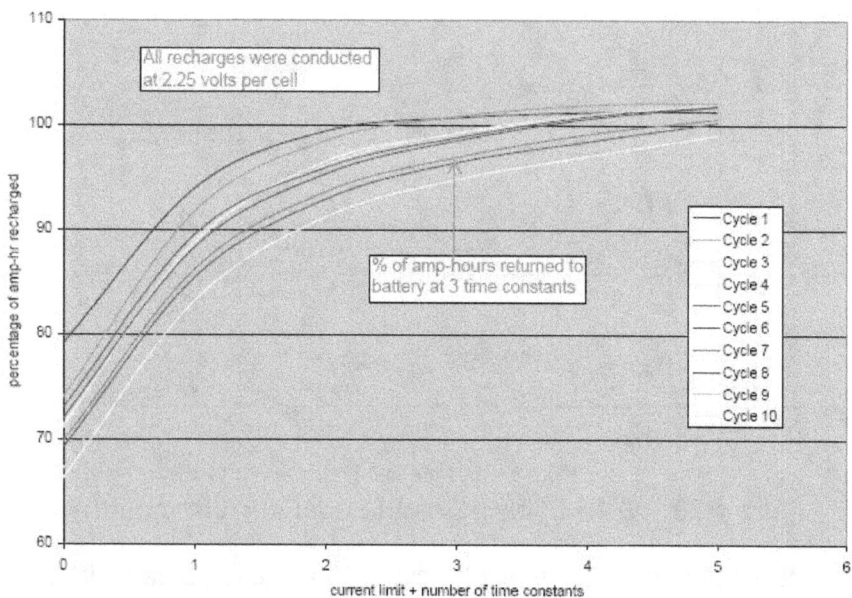

Figure 4-22   Enersys performance during first 10 cycles – ampere hours returned

4.6.2.   GNB "Return-to-service" Tests

"Return-to-service" Testing of the GNB Model NCN-21Batteries was conducted in accordance with the approved test plan. A four hour performance test (discharge test) was conducted followed by a recharge at a float voltage of 27.0 volts. When the battery reached the point where a stable float current was achieved for three hours, a second four hour performance test was conducted. Following completion of the second performance test, the battery was again recharged at a float voltage of 27.0 volts for five days. The process was then repeated a second time. In both tests, the battery was able to meet the performance requirements (>80% of rated capacity). The results of the two "return-to-service" tests are summarized below.

As with the Enersys battery testing, the Alber Battery Capacity Test Set was used to control and monitor the four hour performance tests. It provided data on the discharge current, individual cell voltages, and a real time calculation of battery capacity. The Omega data acquisition system was used to monitor cell temperatures over the entire cycle and charging/float current during the recharge phases of the tests. The Polytronics float current monitor was also used to provide continuous readings of the charging/float current from the battery charger during the recharge cycles. Manual readings of specific gravity were taken periodically during the discharge-recharge cycles.

GNB Cycle 12: The initial performance test for cycle 12 achieved a capacity of 97.0%. The recharge of the battery was initiated and reached a stable float current below 1.0 amps where it remained stable (within 0.5 amps) for the next three hours. At this time the charger output was secured and a series of specific gravity measurements were taken. The second performance test was initiated approximately 22 hours from the start of the recharge. The results from this second capacity test revealed a battery capacity of 95.1%. Following a series of specific gravity measurements, the battery was placed on a 27.0 volt float charge for the remainder of the week. Figure 4-23 illustrates the float current response at the point of stabilization.

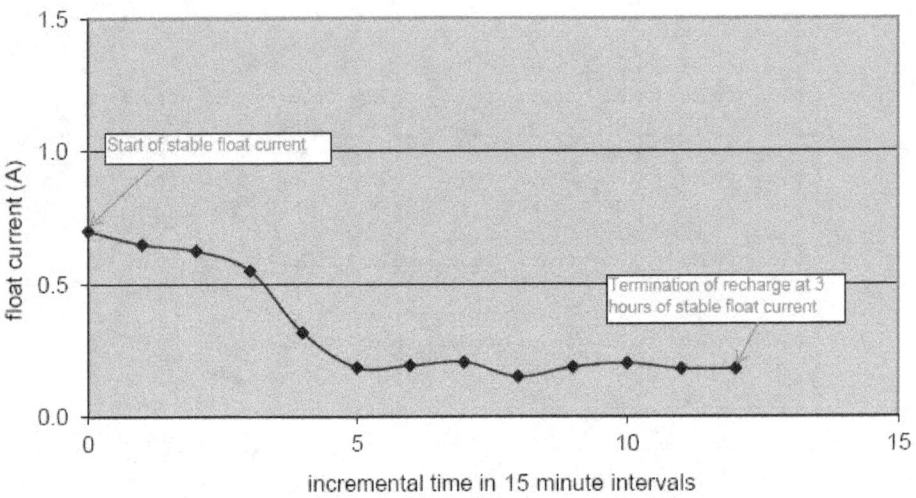

Figure 4-23   GNB Cycle 12 stable float current monitoring

GNB Cycle 13: The initial performance test for cycle 13 achieved a capacity of 95.0%. The recharge of the battery was initiated and reached a float current of about 1.0 amp where it remained stable (within 0.5 amps) for the next three hours. At this time the charger output breaker was opened and a series of specific gravity measurements were taken. The second performance test was initiated approximately 20 hours from the start of the recharge. The results from this second capacity test revealed a battery capacity of 94.4%. Following a series of specific gravity measurements, the battery was placed on a 27.0 volt float charge. Figure 4-24 illustrates the float current curve for GNB cycle 13 used to determine the "return-to-service" point.

Analysis of GNB "return-to-service" testing-stable float current: When the GNB battery was returned to service following a four hour performance test. It was successful in demonstrating that it could meet its performance requirements when recharged to the point where the float current was stable for three hours (GNB cycles 12 and 13). For stable current, we used the criteria that the float current remained the same (within 0.5 amps) for three consecutive hours. In both cases this occurred when the float current was below 1.0 amp. The "Return-to Service" battery capacity achieved in both tests was approximately the same (95.1% in cycle 12 and 94.4% in cycle 13.

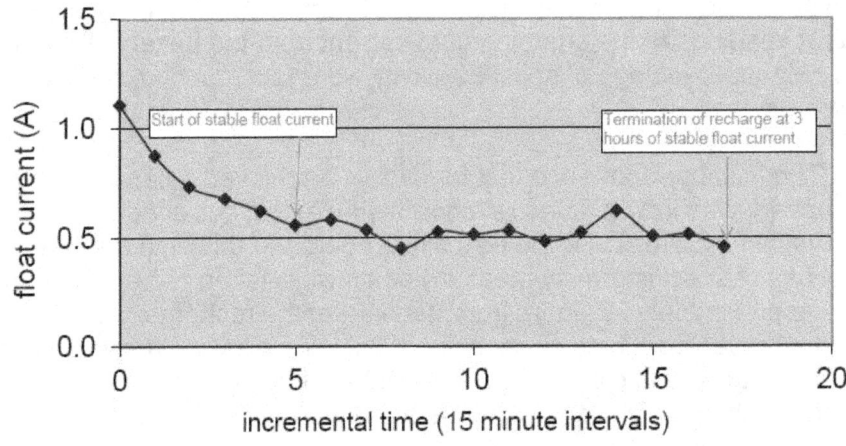

Figure 4-24   GNB cycle 13 stable float current for three hours

64

GNB Cycle 14: "Return-to-service" Testing of the GNB Model NCN-21 battery using a three Time Constant criterion was conducted in accordance with the approved test procedure. The purpose of this test was to confirm that the battery could meet its performance requirements when the float current had reached a point on the exponential float current curve equivalent to three time constants. The float current value at that point was calculated to be 9 amps. To accomplish this, the battery was discharged to a point where the overall string voltage was 21.0 volts. The battery was then recharged at a float voltage of 27.0 volts. When the battery reached the point where the float current was 9 amps, the battery charger was secured and a four hour battery performance test was conducted.

The initial performance test for cycle 14 resulted in a capacity of 94.1%. The recharge of the battery was initiated following the completion of a set of specific gravity readings. The battery reached a float current of 9 amps at approximately 12 hours from the start of the recharge (see Figure 4-25). At this time, the charger output was secured and a series of specific gravity measurements were taken. The second performance test revealed a battery capacity of 94.5%. Following a series of specific gravity measurements, the battery was placed on a 27.0 volt float charge.

Analysis of GNB "Return-to-service" Testing - Three Time Constants: During the first 10 cycles of testing, the calculation of the ampere-hours returned to the battery during recharge indicated that greater than 90% of the ampere hours were returned by the time the float current reached three time constants. Figure 4-26 illustrates this response. Note that cycles 1 and 2, where the recharge was conducted at an equalizing voltage of 28.0 volts (2.33 volts per cell), achieve the 90% level even sooner since the charger stays in a current limit mode longer than when recharge is conducted at a float voltage of 27.0 volts (2.25 volts per cell). In Cycle 14, the four hour performance test conducted at the point where the 3 time constants had been obtained confirmed these results through the use of a performance test which achieved a 94.5% battery capacity, indicative of the substantial ampere-hours returned to the battery at that point during recharge of the battery.

Figure 4-25 illustrates the float current curve following the first discharge test. Highlighted on the curve are the equivalent points for one, two, and three time constants. The charger was secured when the float current reached a value of 9 amps, approximating the three time constant point on the exponential curve. It can therefore be concluded that the battery has the necessary capacity to carry out its design function by the time it is returned to service at the equivalent value of three time constants (9 amps in this case).

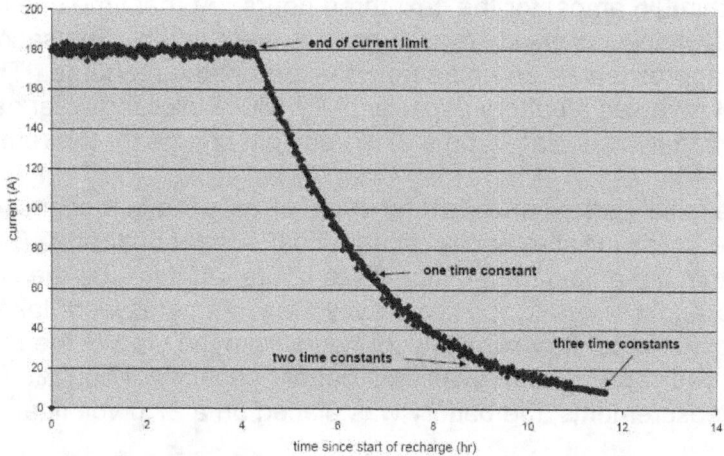

Figure 4-25   GNB cycle 14 float current to three time constants

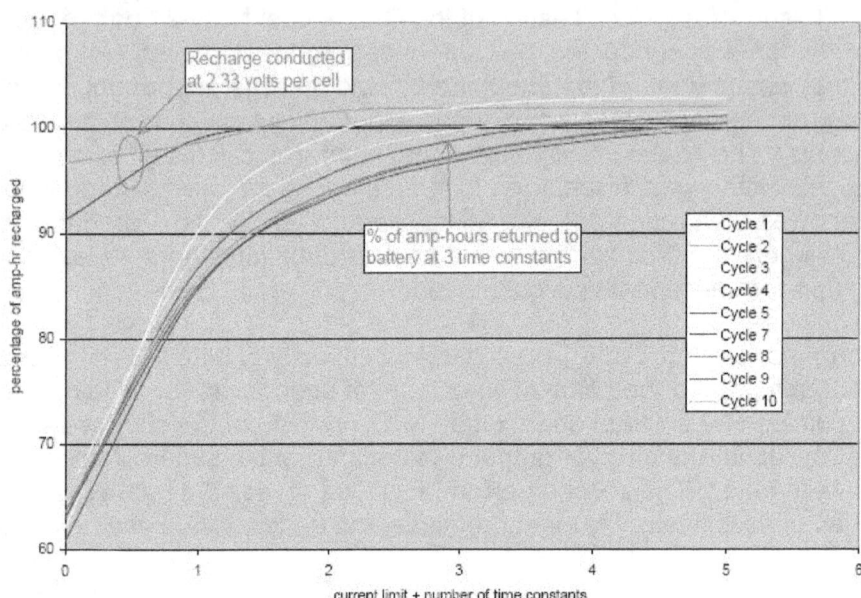

Figure 4-26   GNB float current response during first 10 test cycles

4.6.3.   C&D "Return-to-service" Tests

"Return-to-service" Testing of the C&D Model LCR-33 Batteries was conducted in accordance with the approved test plan. The purpose of this testing was to confirm that the battery could meet its performance requirements when a stable float current was achieved for three hours during recharging of the battery. To accomplish this, a four-hour performance test (discharge test) was conducted followed by a recharge at a float voltage of 27.0 volts. When the battery reached the point where a stable float current was achieved for three hours, a second four hour performance test was conducted. Following completion of the second performance test, the battery was again recharged at a float voltage of 27.0 volts for five days. The process was then repeated a second time (cycle 13). In both tests, the battery was able to meet the performance requirements (>80% of rated capacity). The results of the two "return-to-service tests" are summarized below.

C&D Cycle 12: The initial performance test for cycle 12 resulted in a capacity of 98.3%. The recharge of the battery was initiated and reached a stable float current below 2.0 amps where it remained stable (within 0.5 amps) for the next three hours. At that time, the charger output was secured and a series of specific gravity measurements were taken. The second performance test was initiated at approximately 26 hours from the start of the recharge. The results from this second capacity test revealed a battery capacity of 97.9%. Following a series of specific gravity measurements, the battery was placed on a 27.0 volt float charge for the remainder of the week.

C&D Cycle 13: The initial performance test for cycle 13 resulted in a capacity of 96.6%. The recharge of the battery was initiated and reached a float current of about 2.0 amps where it remained stable (within 0.5 amps) for the next three hours. At this time the charger output breaker was opened and a series of specific gravity measurements were taken. The second performance test was initiated approximately 29 hours from the start of the recharge. The results from this second capacity test revealed a battery capacity of 95.7%. Following a series of specific gravity measurements, the battery was placed on a 27.0 volt float charge.

Analysis of C&D "Return-to-service" Testing - Stable Float Current: When the C&D battery was returned to service following a four-hour performance test, it was successful in

66

demonstrating that it could meet its performance requirements when recharged to the point where the float current was stable for three hours (C&D cycles 12 and 13). For stable current, we used the criterion that the float current remained the same (within 1.0 amp) for three consecutive hours. In both cases this occurred when the float current was about 2.0 amps.

Figures 4-27 and 4-28 are the float current curves for C&D cycles 12 and 13 illustrating the "return-to-service" points.

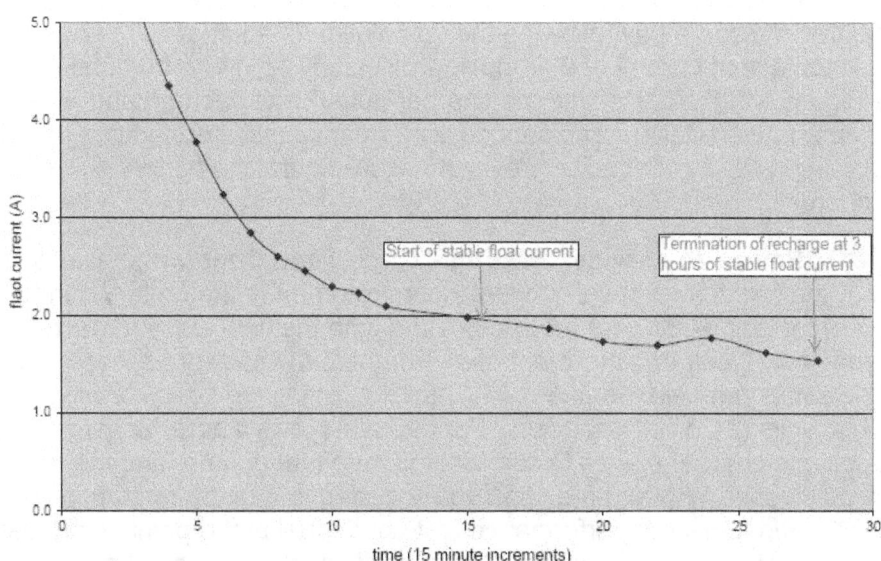

Figure 4-27   C&D cycle 12 stable float current monitoring

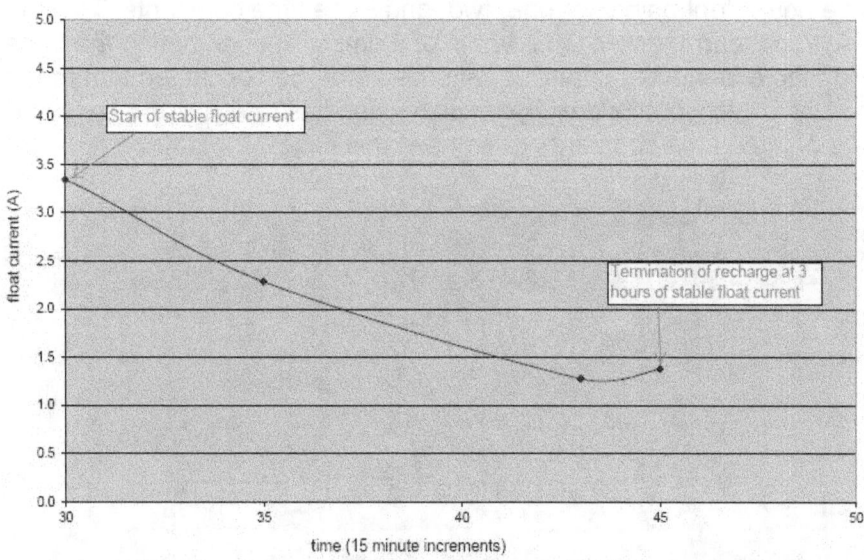

Figure 4-28   C&D cycle 13 stable float current monitoring

C&D Cycle 14: "Return-to-service" Testing of the C&D Model LCR-33 Batteries was conducted in accordance with the approved test procedure. The purpose of this testing was to confirm that the battery could meet its performance requirements when the float current had reached a point on the float current curve equivalent to three time constants. The float current value at that point on the exponential curve is 9 amps. To accomplish this, the battery was discharged to a point where the overall string voltage was 21.0 volts. The battery was then recharged at a float voltage of 27.0 volts. When the battery reached the point where the float current was 9 amps, the battery charger was secured and a four hour battery performance test was conducted. Following completion of the second discharge, the battery was again recharged at a float voltage of 27.0 volts. In this test, the battery was able to meet the performance requirements (>80% of rated capacity). The results of the "return-to-service" test at three time constants" are summarized below.

The initial performance test for cycle 14 resulted in a capacity of 97.9%. The recharge of the battery was initiated the same day following the completion of a set of specific gravity readings. The battery reached a float current of 9 amps approximately 18 hours from the start of the recharge (see Figure 4-29). At this time the charger output was secured and a series of specific gravity measurements were taken. The second performance test resulted in a battery capacity of 96.0%. Following a series of specific gravity measurements, the battery was placed on a 27.0 volt float charge.

Analysis of C&D "Return-to-service" Testing - Three Time Constants: During the first 10 cycles of testing, the calculation of the ampere-hours returned to the battery during recharge indicated that greater than 90% were returned by the time the float current reached 3 time constants. Figure 4-30 illustrates this response. For the C&D battery, recharge was conducted at 28.0 volts (2.33 volts per cell) four times while for the remaining six cycles recharge was conducted at 27.0 volts (2.25 volts per cell). For the four cycles where recharge occurred at 2.33 volts per cell, the charger stayed in current limit for a longer time, and the recharge/float current curve was steeper (shorter time constant) resulting in a faster restoration of ampere-hours. In Cycle 14, the four-hour performance test conducted at the point where the three time constants had been obtained achieved a 96.0% capacity confirming the observation made in Figure 4-30.

Figure 4-29 illustrates the float current curve following the first discharge test. Highlighted on the curve are the equivalent points for one, two, and three time constants. The charger was secured when the float current reached a value of 9 amps, approximating the three time constant point on the exponential curve. It can, therefore, be concluded that the battery has the necessary capacity to carry out its design function when it is returned to service at the equivalent value of three time constants (9 amps in this case).

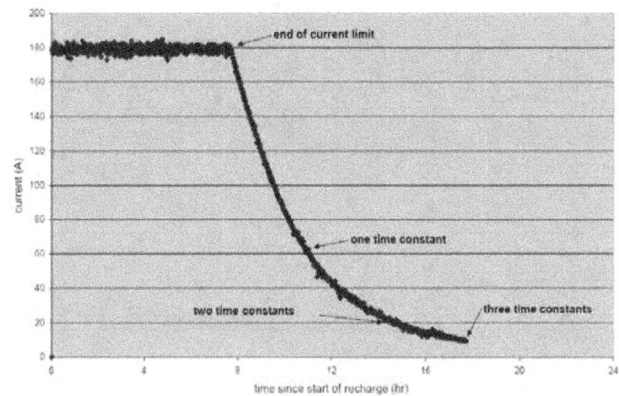

Figure 4-29   C&D Cycle 14 float current response

68

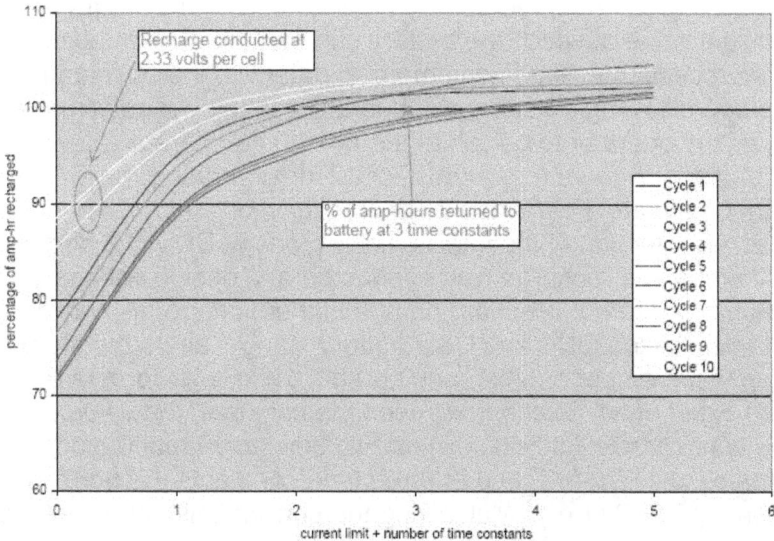

Figure 4-30   C&D float current response during first 10 test cycles

Summary of "Return-to-service" Testing: Table 4-2 summarizes the results of the capacity tests performed for cycles 12-14 for each battery string. The results of this testing indicate that the battery "return-to-service" criteria using stable float current and a float current equivalent to three time constants are both valid means to ensure that a battery can meet its design criteria following its full discharge.

Table 4-2: Summary of "return-to-service" tests

Battery Cycle	Pre-Discharge Capacity	Return to Service Capacity
Enersys 12	101.0%	98.6%
Enersys 13	96.9%	94.8%
Enersys 14	97%*	93.3%
GNB 12	97.0%	95.1%
GNB 13	95.0%	94.4%
GNB 14	94.1%	94.5%
C&D 12	98.3%	97.9%
C&D 13	96.6%	95.7%
C&D 14	97.9%	96.0%

* Estimate by calculation

Cycles 12 & 13 – Return to Service after stable float current for 3 hours
Cycle 14 – Return to Service at 3 time constants (float current = 9 amps)

As noted previously, once the float current reaches an asymptotic level, the ampere-hours being returned to the battery are minimal. This observation relates to the statement in Section C.3 (d) of Regulatory Guide 1.129, Revision 2 that recommends a stable float current reading be achieved for three hours before the battery is considered to be near full charge. The testing performed that is described in Section 4.6 clearly indicates that the battery can meet its performance requirements when a stable float current has been achieved for three hours. In fact, the testing showed that the battery can meet its performance requirements <u>before</u> stable float is achieved (i.e. at the three time constant point on the recharge/float current curve).

The data also showed that for the three battery types tested in this program, the process of selecting a float current value at which the battery can meet its design objectives can be achieved through a process that combines empirical data with the observation that the float current response curve following a battery discharge reasonably approximates an exponential function. This allows the operator to calculate the point (float current value) at which an adequate amount of ampere-hours (e.g., >80% of the design rating) have been returned to the battery to ensure that the battery can meet its design function when returned to service.

For the "return-to-service" testing of the three battery types at BNL, the battery charger current limit was set at 180 amps and recharge was conducted at a design voltage of 2.25 volts per cell or 27.0 volts for the battery string. Recharging at a higher voltage, as was demonstrated in several recharge cycles for the GNB and C&D battery strings, alters the shape of the float current curve (steeper curve/shorter time constant) but the overall form of the curve can still be approximated by an exponential function. At one time constant, the value of the normalized recharge current is approximately 0.368 (1/e), at two time constants the normalized recharge current is approximately 0.135 ($1/e^2$), and at three time constants the normalized recharge current is approximately 0.050 ($1/e^3$). Multiplying the current limit value by 0.050 (three time constants) should yield a current value on the float current curve that represents the point where greater than 80% of the design ampere-hours have been restored to the battery, thereby providing adequate assurance that it can meet its design function assuming that the battery has been properly sized in accordance with IEEE Standard 485-1987.

Because the shape of the recharge curve is not exactly exponential, there will be some variation among battery strings that warrants establishing a battery/battery charger specific recharge current profile to ensure that the approximation described above is reasonable. Table 4.3 illustrates the recharge/float current that actually existed over the 10 cycles of testing. For the Enersys battery, the 9 amp float current value selected using the ideal exponential curve approximates the actual values, ranging from 4.6 to 18.4 amps. For the GNB battery, cycles 1 and 2 were conducted at 28.0 volts so that there is a steeper decline of current with time resulting in values of 1.6 and 1.3 amps, respectively at three time constants. Note that for the remainder of the cycles, the current at 3 time constants ranged from 6.0 to 12.5 amps, a relatively good approximation to the ideal value of 9 amps. Finally, for the C&D battery, the range of currents at 3 time constants is from 3.4 amps to 15.2 amps.

Table 4-3: Actual float current value (Amps) at three time constants

Battery Type	Cycle 1	Cycle 2	Cycle 3	Cycle 4	Cycle 5	Cycle 6	Cycle 7	Cycle 8	Cycle 9	Cycle 10
Enersys	4.6	10.5	14.9	13.7	14.4	15.5	16.6	16.9	16.9	18.4
GNB	1.6*	1.3*	6.0	8.5	10.2	Not obtained	12.3	12.5	12.4	12.5
C&D	3.4	7.1	8.2*	7.4*	13.3	13.3	13.9	15.2	12.8*	9.1*

Recharge conducted at an equalizing voltage of 28.0 volts (2.33 volts per cell)

Using a current shunt similar to the one shown schematically in Figure 2-5 is a convenient and repeatable way to monitor float current during all phases of battery and battery charger operation. While we chose to have the output connected permanently to a data acquisition system, using a calibrated portable digital voltmeter is also acceptable. When performing periodic performance tests, one can not only assess the battery capacity but also monitor the recharge/float current to verify that the float current response curve continues to conform to the general exponential function thereby supporting the use of the three time constant "return-to-service" value. Thus, we conclude that use of the three time constant method is more practical because it provides a current value that can be directly measured.

70

5.   CONCLUSIONS

The primary objective of this research project was to determine whether float current monitoring is a useful indicator for determining a vented lead-calcium battery's state-of-charge. A secondary objective was to evaluate the criteria for selecting the point when a battery can be returned to service and meet its design requirements. In conducting this study, we procured three sets of nuclear qualified batteries from three battery vendors (Enersys, Exide/GNB from NLI, and C&D Technologies) that manufacture nuclear qualified batteries. Each battery set consisted of 12 battery cells. These cells are the same models that are typically used in a Class IE dc system application. Two suitably sized battery chargers and a load bank were also obtained; the second battery charger was used to maintain the batteries not being tested on a continuous float charge.

The test setup was as close as reasonably practicable (not including seismic battery racks) to a typical nuclear power station's Class 1E battery design. Specific gravity measurements were taken in accordance with the battery manufacturer recommendations and IEEE Std. 450-2002. Once the battery was charged and stabilized in accordance with the recommendations in IEEE Std. 450-2002, a series of 4-hour performance discharge tests were performed based on the battery vendor's specifications. After each performance test, the battery was charged and the charging/float current continuously recorded while periodic specific gravity measurements were taken. The specific gravity measurements were taken at the vendor specified location (termed the midpoint) and at the top, midpoint, and bottom of the jar for two of the cells in order to obtain a profile of the electrolyte distribution within the cell. Discharge test current and specific gravity readings were compensated for temperature as discussed in IEEE Std. 450-2002.

From the 30 cycles of testing used to compare specific gravity and float current, we conclude the following with regard to the primary objective of this test program:

1) Both float current and specific gravity provide adequate means to determine battery state-of-charge. Float current has an advantage in that it provides an indicator of the entire battery string, while specific gravity is measured on a cell by cell basis.

2) Both float current and specific gravity have similar response times when the battery is recharged. Generally speaking, 100% of the ampere-hours discharged are returned to the battery within 24 hours of the start of the recharge cycle. Float current response will vary based on the recharge voltage applied to the battery. However, regardless of the voltage applied during recharge, the float current of a nearly fully charged battery becomes stable at less than two amps.

3) The use of pilot cells to ascertain specific gravity is supported by the consistent response observed among all cells during both discharge and recharge.

4) The amount of electrolyte stratification is significant following a performance test and it can take several months before equilibrium is reached again within the cells. Therefore, it is critical to measure specific gravity at the correct distance from the top of the cell as indicated by the battery vendor's manual and supported by IEEE Std. 450-2002. The electrolyte stratification, by itself, does not appear to impact the ability of the battery to meet its capacity and capability requirements.

5) Measuring float current through the use of a simple shunt connected to a data acquisition system provides continuous, accurate and repeatable measurements. A more sophisticated device based on the principles of the Hall Effect (similar to a clamp-on ammeter) was also effective but was less accurate at the low ends of the float current range (< two amps).

6) Recharging at an equalizing voltage rather than a float level voltage results in returning ampere hours to the battery more quickly, but does not consistently improve the time to reach a stable float current (see Table 4-1). The literature suggests that the conditioning of the battery over its life could be detrimentally affected by recharging for extended times at an equalizing voltage. However, no deleterious effects were observed during this testing program under controlled equalizing voltages (not exceeding 40 hours).

Other observations related to the testing of lead calcium batteries used in nuclear power plant applications resulting from this test program are:

Temperature compensation for capacity testing, specific gravity readings, and conductance readings is important and has a measureable impact on the data,
The location of the sampling point (distance from the top of the cell) for the specific gravity measurement is critical due to the significant amount of stratification that occurs following a performance test. Using a standard length of tubing to draw the electrolyte from the same point resulted in consistent "trendable" data, and
Calculations of the ampere-hours returned to the battery during recharge can be used to verify the battery's state-of-charge. The majority (>60%) of the ampere-hours were returned to the battery while the battery charger was in a current limit mode.
IEEE Std. 450-2002 contains the following criterion related to return to service for a battery: "When the charging current has stabilized at the charging voltage for three consecutive hourly measurements, the battery is near full charge." Our test program also verified the point where the battery can be safely returned to service. In a series of six additional tests (two tests per battery string), the battery strings were able to meet their capacity and capability requirements at the point where the float current was stable for three hours. Thus the criterion used in IEEE Std. 450-2002 was found to be an acceptable practice for ensuring the capacity and capability requirements of the battery were met before returning it to service.

Similarly, three cycles of tests were performed in which each battery was returned to service when the float current reached the value equivalent to three time constants on the recharge/float current curve. This occurred within about twelve hours and at a higher current than the previously described return to service tests. In each case, the battery was also able to meet its capacity and capability requirements. This calculated float current value obtained from the battery-specific recharge/float current curve may be a more practical method for returning the battery to service at the point where it is capable of meeting its capacity and capability requirements.

6. REFERENCES

Alber BCT-2000 Software and BCT-128 User's Guide, 4200-003R4.1.1,

Alber, G. and Nispel, M., *Thermal Runaway in Flooded Lead Calcium Batteries*, Tech Note, Alber Corp.

Anderson, James W., *Testing of Large Lead Stationary Batteries, IEEE Transaction on Energy Conversion,* Vol. EC-1, No. 3, September 1986.

C&D Technologies, Inc.; *Standby Battery Vented Cell Installation and Operating Instructions,* RS-1476, Section 12-800, August 2008.

Clark, Steve, *State-of-Charge: Specific Gravity versus Battery Charging Current*, Battcon 2010 Conference Paper.

Determining [conductance] Reference Values," Midtronics, Inc. Website: http://www.midtronics.com/stationary-reference-values-technology/stationary-reference-value-development

DOE Standard 3003-2000, *Backup Power Sources for DOE Facilities*, January 2000.

Enersys Publication No. US-FL-IOM-002, *Safety, Storage, Installation, Operation & Maintenance Manual*, January 2007.

Exide Technologies- GNB Flooded Classic; Installation and Operating Instructions, Section 93.10, March 2010.

Feder, D.O., Croda, T.G., Champlin, K.S., McShane, S.J., and Hlavac, M.J., "Conductance Testing Compared to Traditional Methods of Evaluating the Capacity of Valve-Regulated Lead-Acid Batteries and Predicting State-of-Health," paper presented at the International Lead Zinc Research Organization (ILZRO) Meeting, Nice, France, May 15, 1992.

Feder, D.O., Hlavac, M.J., and Koster, W., "Evaluating the State-of-Health of Lead Acid Flooded and Valve-Regulated Batteries: A Comparison of Conductance Testing vs. Traditional Methods," paper presented at the LABAT '93 International Conference on Lead-Acid Batteries, Varna, Bulgaria, June 7-11, 1993.

Feder, D.O., Hlavac, M.J., and McShane, S.J., "Updated Status of Conductance/Capacity Correlation Studies to Determine the State-of-Health of Automotive SLI and Stand-By Lead Acid Batteries," paper presented at the Zinc and Lead Asian Service (ZALAS) 5th Asian Battery conference 1993, Bali Indonesia, September 1-3, 1993.

Floyd, Kyle et al, *Assessment of Lead-Acid Battery State-of-Charge by Monitoring Float Charging Current*, IEEE *Explore*, 1994.

IEEE Stds. 450-2002 and 2010, *IEEE Recommended Practice for Maintenance, Testing, and Replacement of Vented Lead-Acid Batteries for Stationary Applications.*

Jones, B, "Conductance Monitoring of Recombination Lead Acid Batteries," paper presented at the eleventh International Lead Conference, Venice Italy, May24-27, 1993.

Kniveton, M.W., "Field Experience of Testing VRLA Batteries by Measuring Conductance, paper presented at the 8th Battery Conference and Exhibition, Solihull, Westmidlands, UK, May 11, 1994.

McShane, S.J., and Hlavac, M.J., "Conductance Testing of Standby Batteries in Signaling and Communications Applications for the Purpose of Evaluating Battery State-of-Health," paper presented at the Association of American Railroads' 1993 Eastern Regional Meeting, Lake Buena Vista, Florida, April 29, 1993.

Midtronics Celltron universal stationary battery analyzer, Instruction Manual, 168-641F, April 2006.

Omega Data Acquisition System User's Guide, OMB-DAQ-54/55/56 Data Acquisition Modules, OMB-491-0901, rev 6.1.

Polytronics Engineering Ltd. BtmGlobal System Controller User Manual, Version 3.1.6, 2006.

Regulatory Guide 1.129, *Maintenance, Testing, and Replacement of Vented Lead-Acid Storage Batteries for Nuclear Power Plants,* Rev. 2, 2007.

SBS-2500 Digital Hydrometer, Operating Instructions, November, 2009.

Soileau, Robert D., *A Diagnostic Testing Program for Large Lead Acid Storage Battery Banks,* IEEE Transactions on Industry Applications, Vol. 30, N0. 1, January/February 1994.

Stationary Battery Guide: Design, Application, and Maintenance, EPRI TR- 100248, Rev. 2, August 2002.

Thaller, L. and Zimmerman A., *Understanding and managing capacity walkdown in nickel-hydrogen cells and batteries,* Energy Conversion Engineering Conference, 2002.

APPENDIX A.   THE EXPONENTIAL TIME CONSTANT DURING THE RECHARGE OF A VENTED LEAD-ACID BATTERY

A.1.   DISCUSSION

The state-of-charge of a vented lead-acid storage battery has traditionally been monitored by conducting labor-intensive measurements of the specific gravity of the electrolyte in specified pilot cells, nominally sulfuric acid with a specific gravity of 1.215. However, measurements of the specific gravity of pilot cells during deep discharge and subsequent recharge performance testing have demonstrated that the specific gravity distribution in a battery cell can remain stratified for a long time following the recharge to the fully-charged condition. Therefore, considerable care must be taken in performing specific gravity measurements to ensure that they are always conducted at the same depth into the electrolyte of the pilot cells to ensure consistency between the measurements.

To alleviate the difficulties associated with manual measurements of the specific gravity of selected pilot cells, the IEEE has recently proposed that a battery can be declared fully-charged when the float current during recharge has become stable (not changing with time) for three consecutive hours in recognition of the electrolyte stratification condition, especially following a recharge, an equalizing charge, or addition of water to the cells. The IEEE Standard 450-2002 (p. 7) states that the following may be used as indicators of return to a fully-charged condition following a discharge:

- Stabilized charging current when measured at the manufacturer's recommended voltage and temperature,

- Assurance that the amp-hours returned to the battery are greater than the amp-hours removed plus the charging losses.

Furthermore, IEEE Standard 450-2002 (p. 9) states that, if following a modified performance test the battery delivers a tested capacity of 80% or more (assuming that the battery has been sized in accordance with IEEE Standard 485), the battery is acceptable for service.

IEEE Standard 450-2002 (p. 17) states that "the pattern of charging current delivered by a conventional voltage-regulated charger after a discharge is the most accurate method for determining the state of charge" and, by inference, the most reliable method for determining when the battery can be returned to service.

As a battery is being recharged following a deep discharge, the recharge current will initially be limited to a maximum value set by the charger which we will refer to as the current limit of the charger. As the cells approach full charge, the battery demand will begin to fall below the current limit setting of the charger and will continue to decrease as the voltage of the battery approaches the charger output voltage. The charging current will continue to decrease as the battery capacity (amp-hours) is restored, eventually reaching an asymptotic lower limit at which it will stabilize; at this point, the discharged amp-hours will have been fully restored and the stabilized float current will continue at a very low current to replace the charging losses. A typical example of the current profile vs. time during a recharge cycle following a deep-discharge performance test for a typical battery tested in this program is shown in Figure A-1.

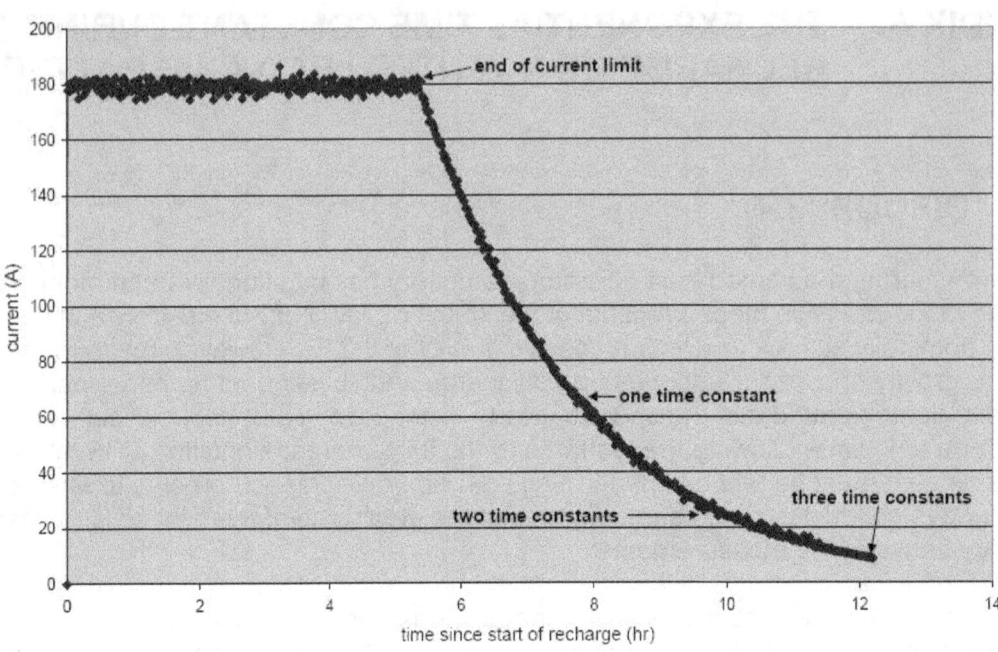

Figure A-1   ENERSYS cycle 14 - float current out to three time constants

In Figure A-1, the time that the battery recharge is in current limit is shown to persist for approximately five hours at 180 amps, after which time the recharge current demand by the battery decreases as the discharged capacity (amp-hours) is restored by the charger. Although not depicted in this Figure (refer to Figure 3-17 for an example), the current would eventually asymptotically approach a lower, stable float current that would usually be in the range less than two amps. The region between the upper current limit and the lower stable float current limit is the region in which the decay of the recharge current vs. time closely resembles a negative exponential function of time. It is in this range that the characteristic exponential time constant for the battery recharge is defined.

The concept of the exponential time constant for the recharge current vs. time in the region between the current limit and the stable float current is a useful metric in considering how long it is necessary to recharge a battery before the battery can be returned to service, even before achieving the fully-charged status as indicated by the three-hour stable float current criterion espoused in IEEE Standard 450-2002 (p. 17).

The first step in calculating the time constant for the exponentially decreasing recharge current is to cull the recharge current data of all data for the current limit period and the stable float current period, so that the remaining data are on the exponentially decreasing portion of the recharge curve. The time axis for the data should be shifted so that the time of the first data point after current limit is set to zero, and the current data should all be normalized by the value of the first data point so that the normalized current data are bounded between unity and zero. The results for this procedure are shown below in Table A-1 for the exponentially decreasing recharge current data for GNB Cycle 8 for some of the data. The data in columns 1 and 2 are the time since start of recharge (which includes the time on current limit) and the measured recharge current in amps. The data in columns 3 and 4 are the time shifted to zero for the first data point after current limit and the recharge current normalized by the value of the first current data point after current limit.

Table A-1   Recharge data for the calculation of the exponential time constant.

Time	Current	Shifted time	Normalized current
(hr)	(A)	(hr)	(-)
4.17	177.4	0.00	1.00
4.18	176.8	0.02	1.00
4.20	175.2	0.03	0.99
4.22	173.8	0.05	0.98
4.23	175.1	0.07	0.97
4.25	170.6	0.08	0.96
4.27	169.9	0.10	0.96
4.28	168.4	0.12	0.95
4.30	167.7	0.13	0.95
4.32	167.8	0.15	0.95
4.33	166.2	0.17	0.94
4.35	164.0	0.18	0.92
4.37	163.3	0.20	0.92
4.38	162.1	0.22	0.91
4.40	162.0	0.23	0.91
4.42	160.1	0.25	0.90
4.43	159.6	0.27	0.90
4.45	158.2	0.28	0.89
4.47	156.8	0.30	0.88
4.48	154.6	0.32	0.87

In this table, only the first 20 data points are listed. In practice, one would use many more data points as shown in Figure A-2 below where the graph includes 200 data points for the exponentially decreasing recharge current for GNB cycle 8.

Assuming that the functional form of the normalized recharge current vs. time data is given by $I/I_0 = K*exp(-bt)$ where K and b are constants, a regression analysis is performed which yields the equation shown in the figure and depicted by the solid line through the data. The constant K is not equal to 1.00 because the data are not perfectly exponential over the range chosen for the curve fit. When the time, t, is equal to one time constant, the value of I/I_0 would be equal to $1/e = 0.36788$. Substituting 0.36788 into the equation in Figure A-2 for I/I_0 and solving the equation for t yields the result for this example that $t_c = 2.30$ hr (see Table 3-5). There are other mathematical approaches to solving for the exponential time constant; however, this example is sufficient to illustrate the underlying rationale.

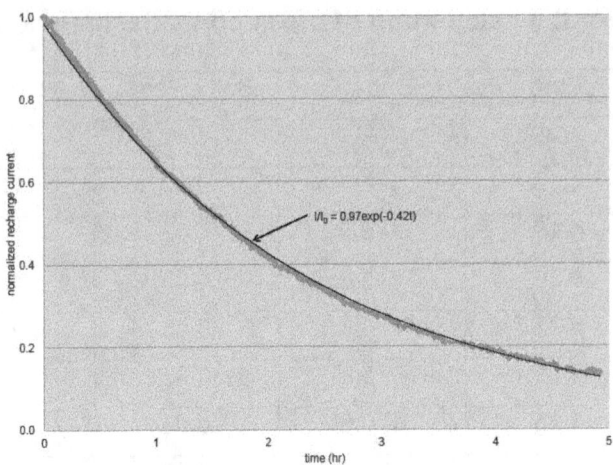

Figure A-2   Normalized recharge current vs. time during the exponentially decreasing recharge current period for GNB cycle 8

An alternative method to solve for the time constant is to realize that the recharge current would have decreased to approximately 0.36788 x 180 amps (the current limit) = 66.2 amps at one time constant. Searching the recorded data for the time after the end of current limit for the recharge current to decrease to 66.2 amps would have yielded a result for the time constant of 2.27 hr, in agreement with the more rigorous approach followed in Table A-1 and Figure A-2.

The utility of the recharge current time constant can be illustrated by consideration of the percentage of the battery capacity that is recharged following a deep-discharge performance test for the three batteries that were tested during this program as shown below in Table A-2. The data for each battery are listed according to test cycle; for each cycle for each battery, the amp-hours discharged during the performance test are listed, followed by the time on current limit during recharge, and the time constant for the exponentially decreasing period of recharge for each cycle for each battery. From Table A-2, it is evident that for all the tests on each of the three batteries tested at least 60% of the battery's capacity was restored during recharge during the time the recharge current was still on current limit. In addition, the data for all the tests indicate that the batteries were recharged to more than 80% of their capacities within one time constant following the departure from current limit, and were recharged to more than 90% of their capacities within two time constants following the departure from current limit.

A-4

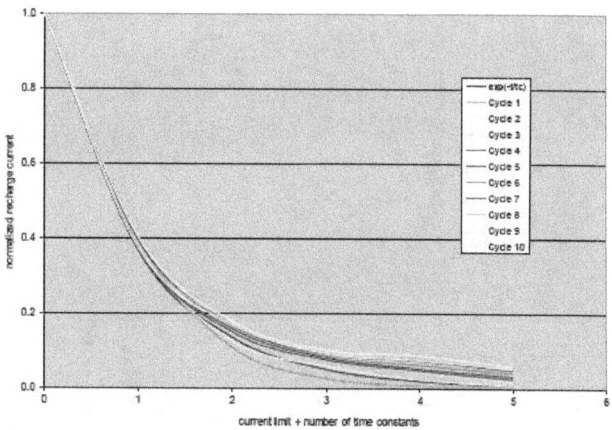

Figure A-3   Normalized recharge current following current limit vs. time constant for ENERSYS cycles 1-10

One can get a better perspective of the dependence of the recharge current and the recharged capacity as a function of time after the current limit period by examining the data in graphical form. For this evaluation, the combined data for the first 10 cycles for the ENERSYS battery are presented below. In Figure A-3, the normalized recharge current is presented vs. time, while Figure A-4 presents the percentage of battery capacity recharged vs. time (both in units of time constants following the current limit period).

It is clear in Figure A-3 that the normalized recharge current closely follows an exponential function out to five time constants beyond the time on current limit. At one time constant, the value of the normalized recharge current should be approximately 0.368 (1/e), at two time constants the normalized recharge current should be approximately 0.135 ($1/e^2$), and at three time constants the normalized recharge current should be approximately 0.050 ($1/e^3$). It is expected that the data will deviate from a true negative exponential with increasing time constants, because the current is not asymptotically approaching zero as would a pure exponential function but is approaching the lower limit of stable float current.

Figure A-4 presents the same data as discussed in Table A-2 but out to five time constants beyond the time on current limit. Here it is evident that the battery capacity rises rapidly until about two time constants beyond the current limit, after which time the rate of recharge decreases, providing little additional capacity afterwards. The utility of the recharge time constant concept in evaluating the state-of-charge and the "return-to-service" limits for a battery is evident from these comparisons. Each battery would have its own unique characteristics and the limits for evaluating the state-of-charge and the "return-to-service" limits for a battery would have to be established on a case-by-case basis. However, the recharge time constant provides an opportunity to reconsider the protocols whereby decisions are made concerning the status of batteries for service.

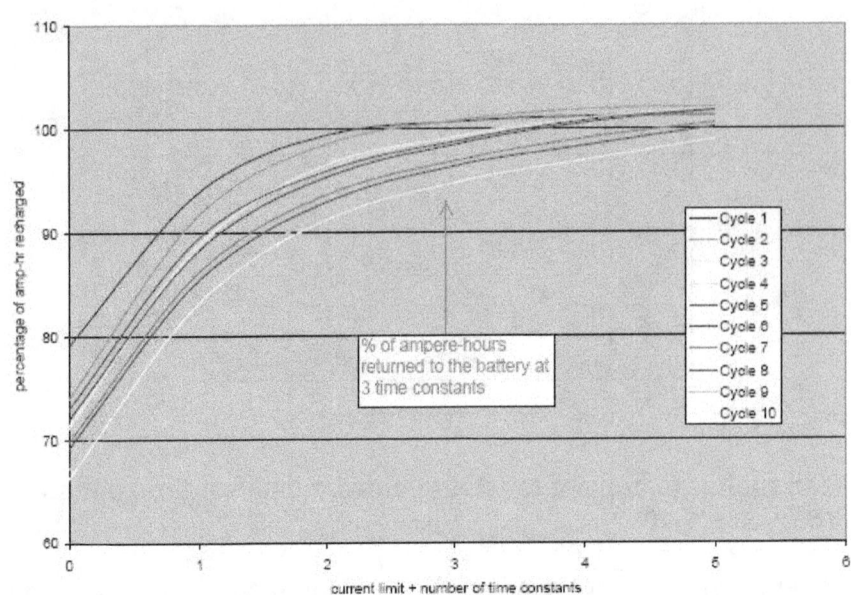

Figure A-4   Percentage of battery capacity recharged following current limit vs. time constant for ENERSYS cycles 1-10

Table A-2   Percentage of the discharged Amp-hours recharged vs. time.

ENERSYS: PERCENT RECHARGE vs. TIME CONSTANT										
Cycle	1	2	3	4	5	6	7	8	9	10
amp-hr discharged	1530	1480	1486	1468	1516	1466	1496	1440	1429	1398
current limit (CL, hr)	6.50	6.00	5.76	5.66	6.06	5.61	5.51	5.28	5.24	4.87
time constant (t_c, hr)	1.98	2.27	2.34	2.38	2.26	2.18	2.24	2.16	2.17	2.16
Percentage of amp-hr recharged that were discharged (all cycles were recharged at 27 volts)										
at current limit	79.1	74.0	71.4	71.1	73.0	72.0	69.8	69.2	67.2	66.3
CL + 1 t_c	93.9	91.7	88.7	89.0	89.5	88.3	86.2	85.4	84.4	83.2
CL + 2 t_c	99.4	98.5	96.5	96.1	95.9	95.3	93.6	92.9	91.5	91.2
CL + 3 t_c	100.7	100.8	99.2	99.3	98.8	98.5	96.9	96.4	95.0	94.8
GNB: PERCENT RECHARGE vs. TIME CONSTANT										
Cycle	1*	2*	3	4	5	6	7	8	9	10
amp-hr discharged	1278	1191	1157	1201	1190		1168	1178	1151	1184
current limit (CL, hr)	6.34	5.67	3.92	4.25	4.15		3.93	4.15	4.03	4.25
time constant (t_c, hr)	0.97	1.22	2.90	2.51	2.51		2.49	2.30	2.28	2.24
Percentage of amp-hr recharged that were discharged (*cycles 1 and 2 were recharged at 28 volts)										
at current limit	91.3	96.7	62.1	64.8	63.0		60.8	63.6	63.1	64.8
CL + 1 t_c	99.0	98.6	90.2	88.3	86.7		84.7	85.6	85.0	85.7
CL + 2 t_c	100.0	101.5	99.3	96.8	95.5		93.6	93.9	93.3	93.6
CL + 3 t_c	100.2	101.8	101.9	99.8	98.9		97.4	97.5	96.9	97.2
C&D: PERCENT RECHARGE vs. TIME CONSTANT										
Cycle	1	2	3*	4*	5	6	7	8	9*	10*
amp-hr discharged	2050	2007	2000	1956	1944	1936	1929	1920	1900	1901
current limit (CL, hr)	8.87	8.52	9.85	9.57	7.82	7.72	7.70	7.60	8.90	9.05
time constant (t_c, hr)	3.03	3.05	1.78	1.80	3.07	3.04	2.98	2.98	2.00	1.94
Percentage of amp-hr recharged that were discharged (*cycles 3, 4, 9 and 10 were recharged at 28 volts)										
at current limit	77.9	76.4	88.5	87.8	73.6	71.8	71.9	71.4	84.3	85.5
CL + 1 t_c	94.9	93.7	98.7	98.5	91.9	89.3	89.0	88.6	96.1	97.0
CL + 2 t_c	100.3	99.8	102.0	102.0	98.9	96.1	95.7	95.2	100.7	101.3
CL + 3 t_c	101.5	101.8	103.0	103.1	101.8	99.1	98.8	98.3	102.7	102.9

NRC FORM 335 (12-2010) NRCMD 3.7	U.S. NUCLEAR REGULATORY COMMISSION **BIBLIOGRAPHIC DATA SHEET** *(See instructions on the reverse)*	1. REPORT NUMBER (Assigned by NRC, Add Vol., Supp., Rev., and Addendum Numbers, if any.) NUREG/CR -7148 BNL-NUREG-98563-2012

2. TITLE AND SUBTITLE

Confirmatory Battery Testing: The Use of Float Current Monitoring to Determine Battery State-of-Charge

3. DATE REPORT PUBLISHED	
MONTH	YEAR
November	2012

4. FIN OR GRANT NUMBER

JCN: 6542

5. AUTHOR(S)

W. Gunther, G. Greene, M. Villaran, Y. Celebi, and J. Higgins

6. TYPE OF REPORT

Technical

7. PERIOD COVERED (Inclusive Dates)

8. PERFORMING ORGANIZATION - NAME AND ADDRESS (If NRC, provide Division, Office or Region, U. S. Nuclear Regulatory Commission, and mailing address; if contractor, provide name and mailing address.)

Energy Science & Technology Department
Brookhaven National Laboratories
Upton, NY 11973-5000

9. SPONSORING ORGANIZATION - NAME AND ADDRESS (If NRC, type "Same as above", if contractor, provide NRC Division, Office or Region, U. S. Nuclear Regulatory Commission, and mailing address.)

Division of Engineering
Office of Nuclear Regulatory Research
U.S. Nuclear Regulatory Commission, Washington DC 205555-0001

10. SUPPLEMENTARY NOTES

L.Ramadan, NRC Project Manager

11. ABSTRACT (200 words or less)

In February 2007, the U.S. Nuclear Regulatory Commission (NRC) issued Regulatory Guide (RG) 1.129 Rev. 2, "Maintenance, Testing, and Replacement of Vented Lead-Acid Storage Batteries for Nuclear Power Plants." In this RG, the NRC staff endorsed the Institute of Electrical and Electronics Engineers (IEEE) Standard 450-2002, "Recommended Practice for Maintenance, Testing, and Replacement of Vented Lead-Acid Batteries for Stationary Applications." This standard provides the recommended practices, test schedules, and testing procedures including recommended methods for determining a battery's state-of-charge to maintain permanently installed vented lead-acid storage batteries (typically of the lead-calcium type) for their standby power applications. Previous versions of this standard suggested that either float current or specific gravity could be used for determining the battery's state-of-charge. The NRC sponsored the research project described herein to validate the use of float charging current as a measure of a battery's state-of-charge for batteries that are used in the nuclear industry. This report describes the approach taken, the specific activities performed to achieve the objectives of this research effort, and the results achieved. It provides analysis of the data and offers observations and recommendations for use by the NRC and its licensees.

12. KEY WORDS/DESCRIPTORS (List words or phrases that will assist researchers in locating the report.)

battery, float charging current, specific gravity, battery capacity and capability, stratification, recharge discharge, and return to service.

13. AVAILABILITY STATEMENT

unlimited

14. SECURITY CLASSIFICATION

(This Page)

unclassified

(This Report)

unclassified

15. NUMBER OF PAGES

16. PRICE

Printed
on recycled
paper

Federal Recycling Program

UNITED STATES
NUCLEAR REGULATORY COMMISSION
WASHINGTON, DC 20555-0001

OFFICIAL BUSINESS

NUREG/CR-7148

Confirmatory Battery Testing: The Use of Float Current Monitoring
to Determine Battery State-of-Charge

November 2012

www.ingramcontent.com/pod-product-compliance
Lightning Source LLC
Chambersburg PA
CBHW080308180526
45167CB00006B/2714